Stefan Güttler

Market assessment of aquaculture research by a small producer country

AF156439

Stefan Güttler

Market assessment of aquaculture research by a small producer country

Südwestdeutscher Verlag für Hochschulschriften

Impressum / Imprint
Bibliografische Information der Deutschen Nationalbibliothek: Die Deutsche Nationalbibliothek verzeichnet diese Publikation in der Deutschen Nationalbibliografie; detaillierte bibliografische Daten sind im Internet über http://dnb.d-nb.de abrufbar.
Alle in diesem Buch genannten Marken und Produktnamen unterliegen warenzeichen-, marken- oder patentrechtlichem Schutz bzw. sind Warenzeichen oder eingetragene Warenzeichen der jeweiligen Inhaber. Die Wiedergabe von Marken, Produktnamen, Gebrauchsnamen, Handelsnamen, Warenbezeichnungen u.s.w. in diesem Werk berechtigt auch ohne besondere Kennzeichnung nicht zu der Annahme, dass solche Namen im Sinne der Warenzeichen- und Markenschutzgesetzgebung als frei zu betrachten wären und daher von jedermann benutzt werden dürften.

Bibliographic information published by the Deutsche Nationalbibliothek: The Deutsche Nationalbibliothek lists this publication in the Deutsche Nationalbibliografie; detailed bibliographic data are available in the Internet at http://dnb.d-nb.de.
Any brand names and product names mentioned in this book are subject to trademark, brand or patent protection and are trademarks or registered trademarks of their respective holders. The use of brand names, product names, common names, trade names, product descriptions etc. even without a particular marking in this works is in no way to be construed to mean that such names may be regarded as unrestricted in respect of trademark and brand protection legislation and could thus be used by anyone.

Coverbild / Cover image: www.ingimage.com

Verlag / Publisher:
Südwestdeutscher Verlag für Hochschulschriften
ist ein Imprint der / is a trademark of
AV Akademikerverlag GmbH & Co. KG
Heinrich-Böcking-Str. 6-8, 66121 Saarbrücken, Deutschland / Germany
Email: info@svh-verlag.de

Herstellung: siehe letzte Seite /
Printed at: see last page
ISBN: 978-3-8381-3681-3

Zugl. / Approved by: Kiel, Christian-Albrechts-Universität zu Kiel, Diss., 2012

Gedruckt mit Genehmigung der
Agrar- und Ernährungswissenschaftlichen Fakultät
der Christian-Albrechts-Universität zu Kiel

Danksagung

Mit der Fertigstellung dieser Dissertation geht ein spannender, lehrreicher und ereignisreicher Lebensabschnitt für mich zu Ende. Im Laufe dieser Zeit haben mich viele Menschen begleitet, bei denen ich mich bedanken möchte.

Meinem Doktorvater Herrn Prof. Dr. Rolf A.E. Müller danke ich für die Überlassung des spannenden Promotionsthemas sowie der Betreuung über all die Jahre. Zudem hat er mich für viele neue Themengebiete, wie z.b. die Welternährung und Informations- und Kommunikationstechnologie, sensibilisiert und meinen Blick auf die Welt geschärft.

Bei Herrn Prof. Dr. Jens-Peter Loy bedanke ich mich für die Übernahme des Zweitgutachtens. Frau Prof. Dr. Nicola Fohrer danke ich für die Übernahme des Vorsitzes der Prüfungskommission und Frau JProf. Dr. Birgit Schulze für ihr Mitwirken in der Prüfungskommission und die gute „Büronachbarschaft".

Herrn Prof. Dr. Carsten Schulz danke ich herzlich für die Unterstützung bei einigen Projekten, wie z.B. der Delphi-Studie und „Fish in the City", der stetigen Hilfsbereitschaft bei Fragen zur Aquakultur und für die Möglichkeit das MASY-Projekt zu begleiten.

Ohne Frau Dr. Susanne Horz hätte ich den Weg der Promotion vermutlich gar nicht erst eingeschlagen. Vielen Dank für den Hinweis auf diese Stelle und die Unterstützung vor, während und nach der Delphi-Studie.

Dank der anfänglichen Einbettung in das FIAF-Projekt war es mir möglich, interessante Forschungsaufenthalte in den USA und Norwegen zu verbringen. Für die freundliche Aufnahme, das Interesse an meinem Forschungsgebiet und die damit verbundene Unterstützung bedanke ich mich bei Herrn Prof. Dr. Julian Alston (UC Davis), Herrn Dr. Stanley Wood und Herrn Dr. Liangzhi You (beide IFPRI, Washington D.C.) und Herrn Prof. Dr. Frank Asche (Universität Stavanger).

Der H.W. Schaumann Stiftung und ihrem Vorsitzenden Herrn Prof. Dr. Dr. h.c. mult. Ernst Kalm danke ich für die finanzielle Unterstützung meiner Konferenzreisen nach Montpellier (IIFET 2010), Zürich (EAAE 2011) und Foz do Iguaçu (IAAE 2012).

Den Forschungsdatenzentren der Statistischen Ämter des Bundes und der Länder danke ich für die Bereitstellung des Datenmaterials für die Schätzung des Nachfragesystems.

Ohne meine Kolleginnen und Kollegen des Instituts für Agrarökonomie der Christian-Albrechts-Universität zu Kiel, besonders aus der Abteilung Innovation und Information sowie dem MultiMediaLabor, hätte die Promotionszeit mir längst nicht so viel Freude bereitet. Herzlich bedanken möchte ich mich bei Franziska Thiemann, Dr. Christiane Ness, Dr. Linda Seidel-Lass, Dr. Doreen Bürgelt, Stephanie Schütze, Christina Bartel und Karsten Borchard für die großartige Unterstützung, den regen Gedankenaustauch und für die tolle Zeit. Zahlreiche wissenschaftliche Hilfskräfte haben ebenfalls zum Fortschritt einzelner Projekte beigetragen. Ein großer Dank geht dabei an Danica Lambrecht, Michaela Haase, Franziska Mierke, Stefan Gelz und an Alexander Schnack.

Meinen Freunden, besonders André Hentschel und Ulf Hennings, möchte ich für die gemeinsame Zeit und die Freizeitaktivitäten außerhalb der Universität sehr danken, aus denen ich immer neue Kraft schöpfen und den Kopf für neue Gedanken frei bekommen konnte.

Ein großer Dank gilt meiner Familie und ganz besonders meinen Eltern für die vielseitige Unterstützung und das entgegengebrachte Verständnis. Mein größter Dank gilt meiner Frau Silvia für die großartige Unterstützung über all die Jahre und die Entlastung in der Endphase der Dissertation.

Stefan Güttler   Kiel, im Januar 2013

STUDIES ON THE MARKET ASSESSMENT OF

AQUACULTURE RESEARCH BY A SMALL PRODUCER COUNTRY

Table of Contents

1. General introduction

1 General introduction

Aquaculture is increasingly important for the future supply of fish because of stagnating supply from fisheries and steadily increasing demand for fish and fish products (FAO 2012a). Aquaculture has a long history and although it has been practiced in China since 475 B.C. (Nash 2011) aquaculture became a significant supplier of fish not before the 1970s (Asche 2008). Thereafter, aquaculture production has increased in the period from 1980 to 2010 by a compound annual growth rate of 8.8 percent and reached 60 mio. t in 2010. At this level of production aquaculture contributes nearly 50 percent to human fish consumption in the world. Due to overfishing and declining fish stocks global fishery production stagnates at a level of about 90 mio. t per year since the mid-1990s. Since then aquaculture has become the sole driver of total fish production growth (FAO 2012a).

From a geographic perspective, Asia is the main driver of aquaculture production growth. About 90 percent of total aquaculture production originates from Asia. Europe's share in total aquaculture production is only about 5 percent. The production share of the EU-27 and EU-15–countries is about 1.7 percent and 1.5 percent, respectively (FAO 2012b). Due to declining fisheries production and increasing domestic demand the European Union became, however, the largest market for imported fish with a share of about 40 percent of total world imports (FAO 2012a).

Recently, the European Commission decided to try to lower the EU's dependency on fish imports by stimulating the domestic production of fish (EU 2009). While fisheries suffer from overexploitation, an increasing aquaculture production may indeed be the best solution for to achieve this objective. Two strategic options are available for increasing aquaculture production. The first is a resource based strategy which involves increasing the resources for aquaculture production. In particular, more fish cages could be installed in the seas, or more fish tanks and ponds could be built on land. However, the potential of a resource-based growth strategy seems limited because available resources of water and land area are limited and often have superior alternative uses (Hilge and Hanel 2008, EU 2009). An alternative strategy to increase the aquaculture production is to shift the aquaculture production frontier by means of research and development (R&D). New and improved technologies may become widely adopted by aquaculture producers resulting in higher productivity, increasing supplies of fish and, potentially, lower market prices. Moreover, new knowledge could also be used as an input for even more new knowledge and inventions so that an accelerating, path-dependent, recursive invention process may emerge in aquaculture (Arthur 2009).

Aquaculture R&D is relatively young and not as highly developed as agricultural livestock research, e.g. research on poultry, pigs, or cattle. The evolution of agriculture has shown that farming organisms that are owned by somebody is much more productive, resource-saving and sustainable compared to hunting and collecting organisms, for which the property rights are not well defined. Compared to fisheries, the property rights are commonly well defined in aquaculture and fish farmers have much better control over the production process than have fishermen (Anderson 2002, Asche 2008). Moreover, although aquaculture R&D is relatively young compared to R&D about traditional livestock species, such as poultry, pigs, or cattle (Asche 2008, Duarte et al. 2007), advances in aquaculture R&D have provided aquaculture farmers with technologies that give them much better control over the production process (Asche 2008). This resulted in an increasing production, a higher productivity, and decreasing production costs, and global aquaculture production started to rise (Asche 2008).

Investments in aquaculture R&D is a means of gaining further technological progress and productivity growth. During the 6th Research Framework Programme (2002 – 2006) the European Commission has invested about 100 mio. € in aquaculture R&D. German states, e.g. Schleswig-Holstein and Bremen, have also launched some sizeable aquaculture-R&D projects to encourage the development of a local aquaculture industry (Seidel-Lass 2009). Aquaculture R&D may generate substantial economic benefits that escape ready measurement and which therefore also escape the attention of R&D policy makers. Public support for aquaculture R&D may be strengthened if it is informed by an *ex ante* analysis of the potential economic benefits of aquaculture R&D.

One aim of this dissertation is to estimate the R&D induced benefits for aquaculture producers and consumers. To do this it is necessary to know the market for fish, and the level and direction of aquaculture R&D. This dissertation therefore consists of several essays, each one dealing with a specific aspect of the estimation of induced benefits from aquaculture R&D. The first essay gives an overview of the development of capture and aquaculture production, as well as fish consumption. The second essay analyses the quality of international fish trade data. The third essay provides a network analysis of international fish trade. The fourth essay deals with the current strength and future direction of aquaculture R&D, as anticipated by aquaculture experts. The fifth essay analyses econometrically the demand for fish in Germany. The last essay, finally, presents a simulation model for estimating the potential benefits of aquaculture R&D. This simulation model makes use of many of the data and insights generated in the earlier essays.

The six essays are now briefly introduced.

The paper **"Captured and cultured fish for food"** serves as a brief introduction and presents the development of capture and aquaculture production and fish consumption in the EU-27. Moreover, the domestication of aquatic organisms is compared to land-based organisms. The development of scientific publications of aquaculture is also compared with publications in fisheries and agriculture.

The European Union depends on fish imports to meet domestic demand. Thus it is necessary to analyze the international trade with fish. Previous studies revealed that trade data are notoriously inaccurate (Morgenstern 1950). Because there is no reason to expect that the quality of trade data for fish is any better than the quality for any other internationally traded good the quality of trade data was scrutinized before the data were used for a network analysis of international fish trade. In the essay, entitled **"Testing the quality of international fish trade data"**, the import and export data of three fish product categories are analyzed for the period from 1992 to 2008. In a first step it is examined whether the trade data follow Benford's law. Benford's law has been applied to detect accountancy and tax fraud (Nigrini 1996, 1999) or to check the reliability of survey data (Judge and Schechter 2009, Schräpler 2010). In this study it is applied to import and export data of fish. Further, bilateral trade data are scrutinized. The double-recording of a trade flow as an export by the exporting country and as an import by the importing country, offers the possibility to compare and judge trade statistics. This analysis provides the possibility to identify countries which may systematically over- or understate their international fish trade.

In the third essay **"Fish in the network – network analysis of international fish trade"** the analysis of the fish market is extended by examining international fish trade with methods of network analysis. Networks are based on graph theory and international trade can be represented as a graph with the countries as nodes and the trade flows as (weighted) arcs connecting the nodes. Trade networks of three different fish species have been scrutinized for the period from 1990 to 2009. The development and patterns of fish trade are checked whether there are differences in the networks for fish originating from aquaculture or from capture fisheries. The main importing and exporting countries are identified in terms of trade value as well as for the number of trade connections. Moreover, countries are identified which play an important role as intermediaries in the fish supply chain.

A study of the future impacts of R&D should provide some information on the current state of aquaculture R&D as well as the future direction of aquaculture R&D. The fourth essay **"The shape of future aquaculture R&D – results of a Delphi study"** deals with this issue. Aquaculture experts were surveyed in an international Delphi study comprising three survey rounds. Aquaculture R&D in high-income countries was rated by the experts. The aquaculture experts also rated the current strength and future development of several R&D fields, e.g. fish breeding and reproduction, fish nutrition, and fish husbandry and water management. The Delphi study identified the most promising research areas and could also help funding agencies and decision makers to identify promising areas of aquaculture R&D.

An important goal of applied food commodity research always is to reduce production costs and some aquaculture R&D activities can be expected to lead to a reduction in fish production costs. Producers would certainly benefit from lower production costs but so would consumers if lower production costs would also translate into lower market prices which result when the supply curve is shifted downwards by means of R&D. To estimate the economic benefits of R&D for consumers it is important to know how consumers react to price changes. In the fifth essay **"Demand for fish in Germany"** a quadratic almost ideal demand system (QUAIDS) is estimated for Germany and own-price elasticities for fish and several fish products are estimated. Missing price information and zero observations in the data set may lead to biased results. Two methods were applied to control for these effects, namely the consistent-two-step estimation and quality adjusted prices. A sensitivity analysis was conducted to control for the effect of these two methods on the estimation results and price elasticities.

Finally, in the essay **"Simulating the benefits from aquaculture R&D – the impact of elasticities and spillovers"** the results of the demand analysis are used in the estimation of the potential welfare effects induced by aquaculture R&D. A simulation model is built for the EU-15 countries to measure the impact on consumer and producer surplus of aquaculture R&D conducted in Germany. As the results depend on many parameters the influence of income elasticities, demand elasticities, supply elasticities, and spillover coefficients on the benefits is examined. Simulations are run for four different spillover matrices. In one of these scenarios the spillovers are based on a bibliometric study of scientific publications in aquaculture and fisheries (Seidel-Lass 2009). This essay is an extension of previous papers (Guettler et al. 2010, 2012), where the influence of adoption lags and research lags on total benefits has been analyzed. The results provide important implications for policy decisions concerning the allocation of public funds for aquaculture R&D-projects.

Subsequent to the sixth essay the main results of all papers are summarized.

This thesis provides new insights on the market for fish, as the demand in Germany and the international fish trade are analyzed. Additionally, the future direction of aquaculture R&D in high-income countries is identified. The economic evaluation of aquaculture R&D conducted in Germany provides a basis for decision making and public investments into aquaculture R&D.

References

Anderson, J.L. (2002): Aquaculture and the future: why fisheries economists should care. Marine Resource Economics 17(2): 133-151.

Arthur, W.B. (2009): The nature of technology. Free Press, New York.

Asche, F. (2008): Farming the sea. Marine Resource Economics 23(4): 527-547.

Duarte, C.M., Marbá, N. and Holmer, M. (2007): Rapid domestication of marine species. Science 316: 382-383.

EU (2009): Building a sustainable future for aquaculture: A new impetus for the Strategy for Sustainable Development of European Aquaculture, Communication from the Commission to the European Parliament and the Council, COM (2009) 162, Brussels.

FAO (2012a): The state of world fisheries and aquaculture 2012. Food and Agriculture Organization of the United Nations, Rome.

FAO (2012b): Fisheries and Aquaculture Department, Statistics and Information Service FishStatJ: Universal software for fishery statistical time series. <http://www.fao.org/fishery/statistics/software/fishstatj/en>.

Guettler, S. and Mueller, R.A.E. (2010): Benefits from R&D and spill-overs in aquaculture: An EU-15 modelling approach. In: Proceedings of the Fifteenth Biennial Conference of the International Institute of Fisheries Economics & Trade, July 13-16, 2010, Montpellier, France: Economics of Fish Resources and Aquatic Ecosystems: Balancing Uses, Balancing Costs. Compiled by Ann L. Shriver. International Institute of Fisheries Economics & Trade, Corvallis, Oregon, USA, 2010. CD ROM. ISBN 0-9763432-6-6.

Guettler, S., Seidel-Lass, L. and Mueller, R.A.E. (2012): Simulating the spillover benefits from R&D by a small producer country embedded in a co-authorship network: Aquaculture R&D in Germany. Selected Paper, IAAE 2012 Conference, August 18-24, 2012, Foz do Iguaçu, Brazil. <http://ageconsearch.umn.edu/handle/122885>

Hilge, V. and Hanel, R. (2008): Aquakultur: bedeutend für die Welternährung. ForschungsReport 2: 11-13. Federal Ministry of Food, Agriculture and Consumer Protection, Bonn, Berlin.

Judge, G. and Schechter, L. (2009): Detecting problems in survey data using Benford's law. Journal of Human Resources 44 (1): 1-24.

Morgenstern, O. (1950): On the accuracy of economic observations. Princeton University Press, Princeton.

Nash, C.E. (2011): The history of aquaculture. Wiley-Blackwell, Ames, IO.

Nigrini, M.J. (1996): A taxpayer compliance application of Benford's law. The Journal of the American Taxation Association 18: 72–91.

Nigrini, M.J. (1999): Adding value with digital analysis. The Internal Auditor 56: 21-23.

Schräpler, J.P. (2010): Benford's law as an instrument for fraud detection in surveys using the data of the Socio-Economic Panel (SOEP). SOEPpapers on Multidisciplinary Panel Data Research at DIW Berlin 273, February 2010, Berlin. <http://www.diw.de/documents/publikationen/73/diw_01.c.349061.de/diw _sp0273.pdf>.

Seidel-Lass, L. (2009): Networks in international aquaculture research: a bibliometric analysis. Cuvillier Verlag, Goettingen.

2. Captured and cultured fish for food

Stefan Guettler, Danica Lambrecht and Rolf A.E. Mueller

Published in: EuroChoices 11(1), 2012, pp. 26-27.

http://onlinelibrary.wiley.com/doi/10.1111/j.1746-692X.2012.00221.x/abstract

2. Captured and cultured fish for food

The shift from hunting and gathering to agriculture which occurred some 10,000 years ago involved the purposeful control and modification of terrestrial plants and animals. Fish, however, largely escaped a similar domestication at the dawn of agriculture. This omission from 10,000 years ago has been remedied over the last 100 years and fish have been domesticated at a rapid rate (Figure 1). For instance, nearly 70 per cent of all marine animals domesticated up to the present have been domesticated in the past 30 years.

Figure 1: Progress of domestication of animals and plants, 10,000 years ago to present

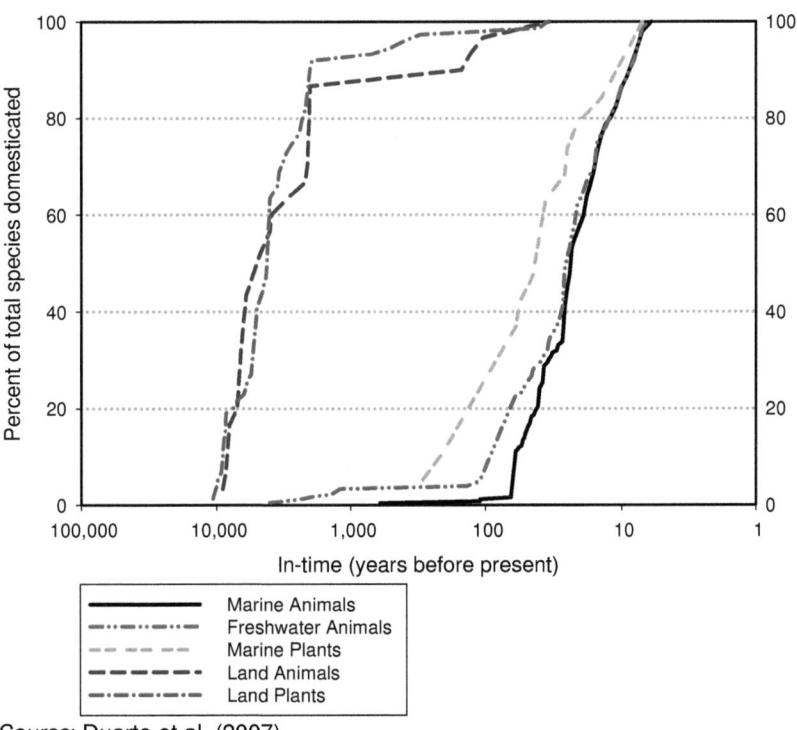

Source: Duarte et al. (2007)

Total world finfish production (vertebrate and cartilaginous fish species, excluding crustaceans, cephalopds, and molluscs) has grown from 17 to 112 million tonnes between 1950 and 2009 (Figure 2). The share of cultured fish in total world fish production has been increasing from less than 2 per cent in 1950 to 32 per cent in 2009. Moreover, fisheries experts suggest that all future increases in demand for fish for food will have to be met by cultured fish. China has emerged as a major aquaculture producer. In comparison to China, the EU is a minor aquaculture producer.

Figure 2: Development of capture and aquaculture finfish production, 1950-2009 [mio. t.]

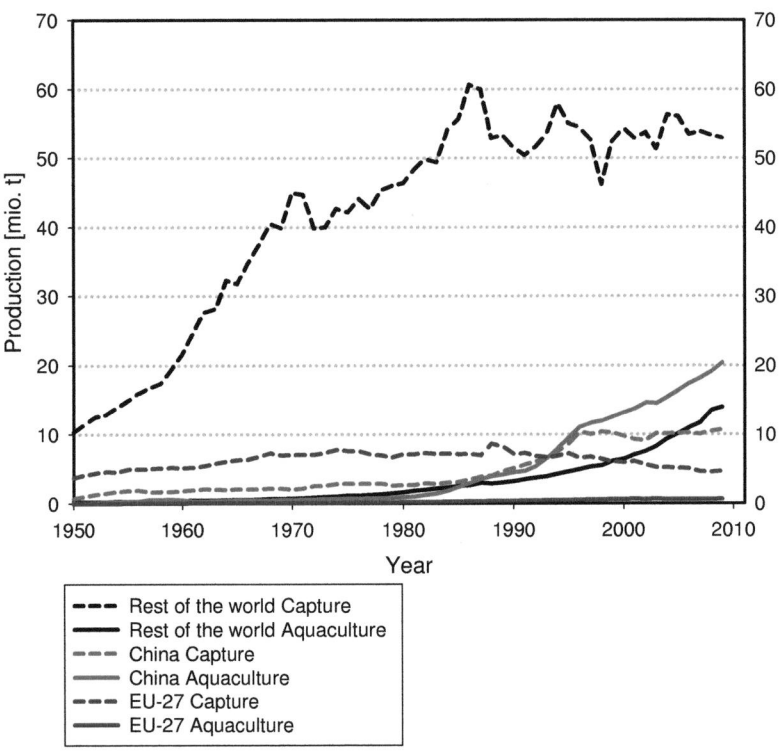

Source: FAO (2011a)

Per capita fish consumption in the EU has grown by 0.5 per cent per year from 13.2 to 16.4 kg/yr between 1961 and 2007 (Figure 3). The variation of per capita fish consumption in the EU-27 is considerable: from 4 kg/p.c./yr in Bulgaria to 45 kg/p.c./yr in Portugal (Figure 4). Unfortunately, available data do not allow us to distinguish between consumption of captured and cultured fish.

Fish prices have remained at similar levels to the prices of major substitutes. In the EU-27 the Harmonized Consumer Price Index (2005 =100) for fish stood at 73 in 1996 and increased to 113.3 in 2010, while the index for meat increased from 78.8 in 1996 to 112.8 in 2010.

Figure 3: Annual per capita consumption of finfish, 1961-2007 [kg/capita]

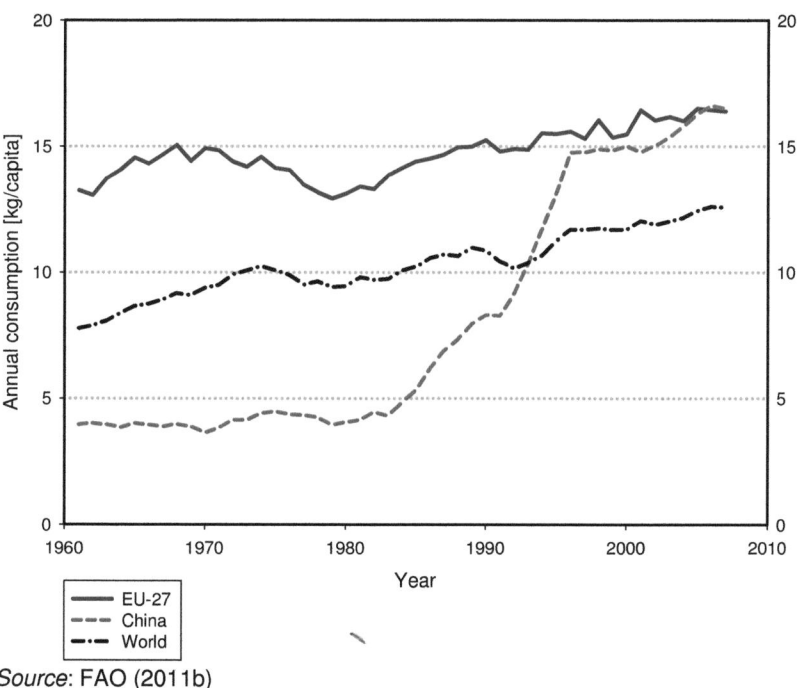

Source: FAO (2011b)

Figure 4: Consumption of meat and fish in Europe, 2007 [kg/capita]

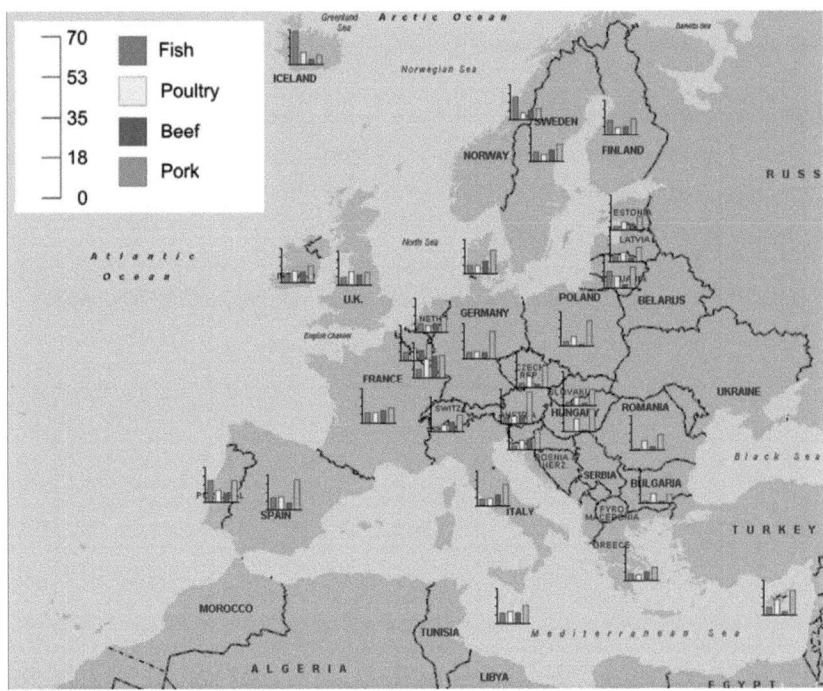

Source: FAO (2011b)

The large and increasing gap between domestic fish production and consumption in the EU-27 is balanced by increasing net imports of fish (Figure 5). Important exporting countries of fish to the EU-27 currently are Norway and China.

Figure 5: Development of net imports and EU-27 aquaculture production, 1988-2010

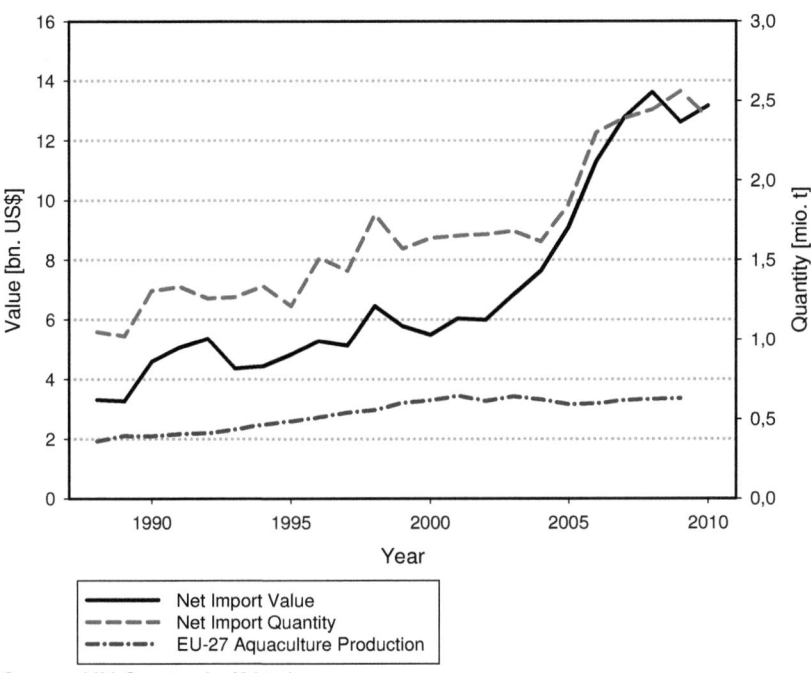

Source: UN Comtrade (2011)

The development of aquaculture is driven by R&D. A useful indicator for R&D intensity is the number of scientific publications registered by ISI's Science Citation Index (Figure 6). The development of the number of publications by subject category suggests that aquaculture is no longer a field neglected by R&D. If this trend in aquaculture R&D is sustained there is hope that the size and productivity of aquaculture will grow within a few decades to a level where cultured fish is able to substitute for captured fish.

Figure 6: Development of scientific publications by research area, 1951-2010

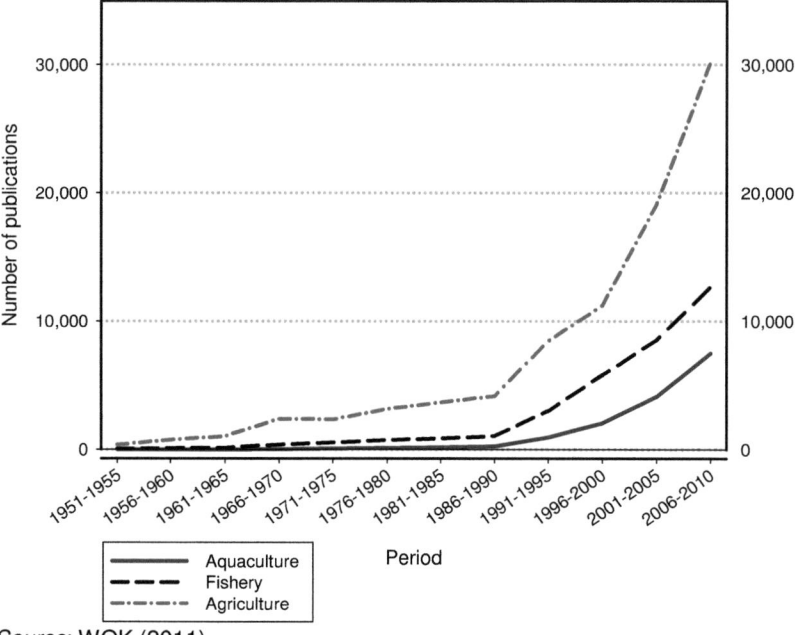

Source: WOK (2011)

References

Duarte, C.M., Marbá, N. and Holmer, M. (2007): Rapid domestication of marine species. Science, 316, pp. 382-383.

FAO (2011a): FAO Fisheries and Aquaculture Department, Statistics and Information Service, FishStatJ: Universal software for fishery statistical time series. <http://www.fao.org/fishery/statistics/software/fishstatj/en>.

FAO (2011b): FAOSTAT <http://faostat.fao.org/>.

UN Comtrade (2011): United Nations Commodity Trade Statistics Database. <http://comtrade.un.org/db/default.aspx>.

WOK (2011): Web of Knowledge <http://www.webofknowledge.com>.

3. Testing the quality of international fish trade data

3 Testing the quality of international fish trade data

Abstract

Many economic and econometric publications deal with trade data, but the quality of the trade data is rarely examined. Probably one of the first studies dealing with the accuracy of data was Morgensterns` book published in 1950. Morgenstern (1950) detected serious inaccuracies in trade data and more recent studies by Yeats (1990) and Rozanski and Yeats (1994) reveal significant deficiencies in trade statistics.

The aim of this paper is to analyze the quality of fish trade data in the UN Comtrade data base. To this end two tests to the fish trade data are applied. First, it is examined if the first digits of the reported trade values follow a special distribution, viz. Benford's law. In a second step bilateral import and export data are used to calculate percentage differences and classify them into four categories suggested by Morgenstern. Concentration indices calculated on the distribution of these four cases give a hint whether data may be biased.

Both tests are applied to several commodity classification codes, to check (i) whether there are differences between them and (ii) whether there are differences in the trade data between fish originating mainly from fisheries or aquaculture.

Results show that trade data from several countries violate Benford's law suggesting deficiencies in data quality. A country by country comparison of trade data revealed large discrepancies between import and export data and, depending on the commodity classification, 50 percent to 80 percent of the import and export data differ by more than ±50 percent from each other. However, systematic pattern in the discrepancies between fish export and import data could not be detected. The results suggest that fish trade data should be scrutinized for each country and year before they are turned into information by interpretation or quantitative models.

1   Introduction

International trade with fish strongly increased in the period from 1976 to 2008 with a compound annual growth rate of 8 percent in value terms (FAO 2010). In 2008, trade in fish and fish products accounted for 10 percent of total agricultural exports or 1 percent of world merchandise trade. Drivers of the expansion of fish trade were changes in the fishery sector, globalization of the fisheries and aquaculture value chain and of food processing systems, increasing consumption, trade liberalization, and various technological innovations, e.g. in packaging, processing, and transportation (FAO 2010).

In the last decade global capture fishery production stagnated and the increase of aquaculture production alone was responsible for an increasing supply of fish. Today, nearly 50 percent of human fish consumption is supplied by aquaculture (FAO 2010). It is therefore not surprising that products derived from aquaculture production are contributing an increasing share of international fish trade. As the classifications used by international trade statistics do not distinguish between fish from fisheries and fish from aquaculture, it is difficult to determine the share of aquaculture in total fish trade (FAO 2010). Nevertheless, one of the fish products analyzed in this study is salmon, which is mainly produced in aquaculture.

Already in 1950 Morgenstern detected discrepancies in bilateral trade data. Trade data are recorded twice, once by the exporting country and once by the importing country, although it may be not always clear which country was the first exporter and different valuation of import and export flows may lead to differences between the recorded import and export data. However, the similar event is observed by two observers (exporting and importing country). Records of similar events therefore provide a basis for the judgment of the quality of trade data. This is not the case for other, e.g. purely domestic data like production data, which are recorded only once. Since Morgenstern's book was first published in 1950, recording and processing of trade data has changed and many quality programs have been implemented by statistical agencies (UN 2004, Bergdahl et al. 2007). Nevertheless, the quality of international trade data still seems to need improvements, because more recent publications by Yeats

(1990), Rozanski and Yeats (1994), Martin (2010) and others revealed discrepancies in trade data and thus confirm the findings by Morgenstern.

The quality of data is of great practical interest, especially when trade policy decisions or economic models are based on them. It is indisputable that incorrect data used in other studies, other economic trade models, etc. may lead to biased results and suggestions. But, trade data are always imperfect: random errors and systematic errors or biases lead to deviations between the observed value and the "true value". Random errors have two attributes that render them benign in comparison with biases. First, the law of large numbers assures that positive and negative deviations of measured values from true values balance out when the number of observations is sufficiently large. Second, the size of random errors can be measured. Not so with systematic errors: they don't balance out but accumulate and there is no measure for their size.

The goal of this paper is to scrutinize the quality of fish trade data by several commodity codes to examine whether differences in the data quality between various commodities exist. The outline is as follows: First the data source and commodity classification systems are described. The third section reviews the development of fish trade and the main trading countries. A first quality test is done by examining if the first significant digit of trade data follows Benford's law. Morgenstern's classification is applied to trade data of three commodity classes to gain more information on the type of discrepancies. Both tests try to identify countries whose trade data are biased. The last section discusses the results.

2 Data selection

2.1 Selection of data source

International fish trade data are collected and published by the Organization of Economic Co-operation and Development (OECD), the Statistical Office of the European Union (Eurostat) and the United Nations (UN). For a comparison of bilateral trade data the UN Comtrade (http://comtrade.un.org) data base was chosen for this paper as it is the only data source which provides data for trade between pair of countries. This comprehensive database for international merchandise trade statistics contains about 1.7 billion trade records in 9 classifications up to the 6-digit level of the classification. Data are available in the three classifications, viz. (i) Harmonized Commodity Description and Coding System (Harmonized System, or HS), (ii) Standard International Trade Classification (SITC), and (iii) Broad Economic Categories (BEC).

2.2 Selection of commodity classification

For this study the SITC Revision 3 was chosen, which covers a longer period than the current SITC Revision 4, which was promulgated in 2006. Furthermore, the SITC classification is recommended for analytical purposes by UN (2008) and thus the HS classifications was not considered in this study. The BEC classification covers only highly aggregated commodity groups and is therfore not suitable for the comparison of different fish products.

The SITC has a hierarchical structure:

Sections	1-digit code;
Divisions	2-digit code;
Groups	3-digit code;
Subgroups	4-digit code;
Items	5-digit code.

Section 0 "Food and live animals" includes fish, crustaceans, molluscs and aquatic invertebrates in division 03. Division 03 is further divided into four groups 034-037, some of which are further divided into subgroups (see Table 1).

International trade classifications do not differentiate between the origin of a fish or fish product. In particular, trade statistics are mute about whether a fish originates from fishery or aquaculture. Only some fish species like tuna, salmonidae, flat fish, cod, hake, herring and mackerel are explicitly identified. Of these species only salmonidae are predominantly produced in aquaculture with a share of 75 percent of total world production in the year 2008 (FAO 2011). It is for this reason that the SITC-codes 03412, 03421 and 03711 are of special interest for the following analysis.

As SITC Rev. 4 reports only data from 2007 and later, the trade data from SITC Rev. 3 were chosen. A quick check of data availability at UN Comtrade's data base for SITC Rev. 3 showed that data availability is poor for many countries before 1992. Moreover, data for the years 2009 and 2010 were not fully reported by all countries. Therefore only data for the period from 1992 to 2008 were included in this study.

Table 1: Description of selected SITC-Codes

SITC-Code	Description
03	Fish (not marine mammals), crustaceans, molluscs and aquatic invertebrates, and preparations thereof
034	Fish fresh (live or dead), chilled or frozen
03412	Salmonidae, fresh or chilled (excluding livers and roes)
03421	Salmonidae, frozen (excluding livers and roes)
035	Fish, dried, salted or in brine; smoked fish (whether or not cooked before or during the smoking process)
036	Crustaceans, molluscs and aquatic invertebrates, whether in shell or not, fresh (live or dead), chilled, frozen, dried, salted or in brine; crustaceans, in shell, cooked by steaming or boiling in water
037	Fish, crustaceans, molluscs and other aquatic invertebrates, prepared or preserved, n.e.s.
03711	Salmon, whole or in pieces, but not minced

Source: UN (1991)

3 International fish trade

3.1 Development of fish trade

International trade with fish increased strongly in the last years. World fish exports nearly trippled from 33 bn. US$ in 1992 to 89 bn. US$ in 2008 and world imports increased from 41 bn. US$ to nearly 100 bn. US$ in the same period (see Figure 1). Noteworthy is the strong increase between 2002 and 2008: since 2002 fish exports increased from 51 bn. US$ to 89 bn. US$ in 2008. This is equivalent to a compound annual growth rate of 9.7 percent. However, it is estimated that fish exports decreased by 7 percent worldwide due to the economic crisis of 2008 (FAO 2010). However, fish and fish products are highly traded and trade grew significantly in the last decades.

Of course, developments of world exports and imports of fish show the same pattern, but it can also be seen in Figure 1 that imports are on average valued 15 percent higher than exports with a minimum of +9.8 percent in 2006 and a maximum of +26.3 percent in 1992. An unknown share of the deviation between import and export figures may be explained by the fact that imports are mainly valued on cost-insurance-freight (cif) basis while exports are valued free on board (fob) (UN 2004).

Figure 1: Development of total world imports and exports of fish, crustaceans, molluscs and aquatic invertebrates, and preparations thereof (SITC-code 03), 1992 – 2008 in bn. US$

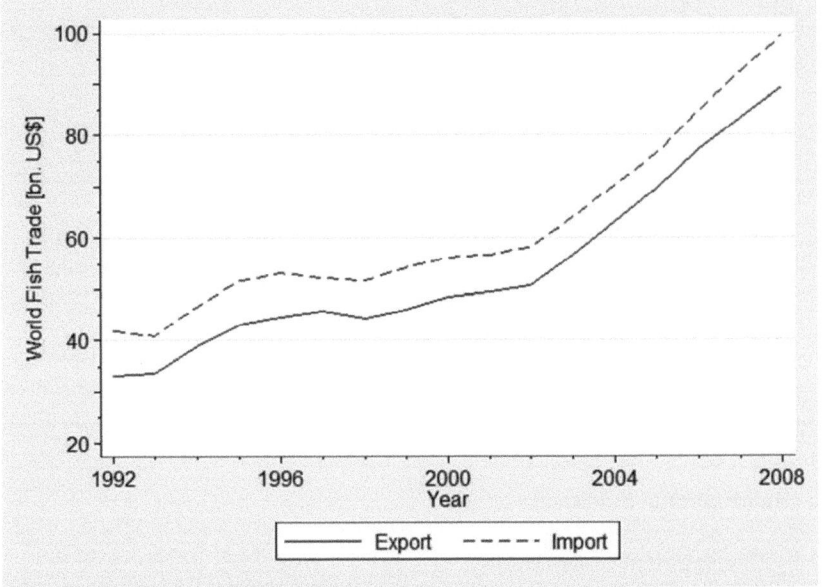

Data source: UN Comtrade (2010)

The rise in total fish trade is mainly due to increase in fresh, chilled or frozen fish (SITC-code 034), whose exports in value terms rose from 14 bn. US$ to 42 bn. US$ between 1992 and 2008 (see Figure 2). Trade with crustaceans, molluscs and other aquatic invertebrates (SITC-code 036 and 037) also increased remarkably and accounted for a trade volume of 21 bn. US$ in 2008. Trade with dried, salted or smoked fish (SITC-code 035) in contrast accounted for less than 5 bn. US$ and has increased only moderately since 2004.

Salmonidae is the only product category which is predominantly produced by aquaculture which can be identified in trade statistics becaude it is given an own commodity code. Trade with fresh, chilled salmonidae (SITC-code 03412) increased by a compound annual growth rate of 15.6 percent between 2001 and 2008 and reached 5.4 bn. US$ in 2008 (see Figure 2). Trade with prepared or preserved salmon (SITC-code 03711) is nearly insignificant and trade with frozen salmon (SITC-code 03421) increased in the last years and stood at 2.4 bn. US$ in 2008.

Figure 2: Development of total world fish exports by SITC groups, 1992 – 2008 in bn. US$

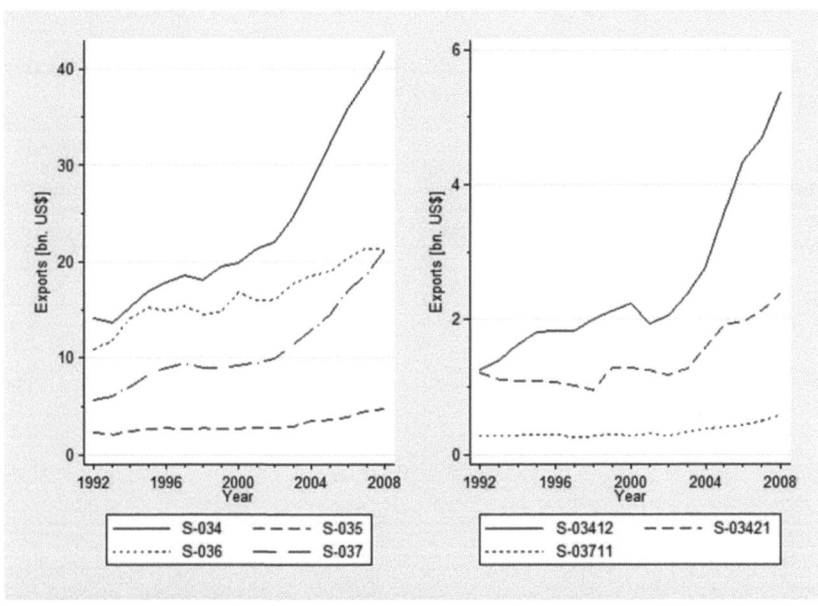

S-034:	Fish fresh (live or dead), chilled or frozen
S-035:	Fish, dried, salted or in brine; smoked fish (whether or not cooked before or during the smoking process)
S-036:	Crustaceans, molluscs and aquatic invertebrates, whether in shell or not, fresh (live or dead), chilled, frozen, dried, salted or in brine; crustaceans, in shell, cooked by steaming or boiling in water
S-037:	Fish, crustaceans, molluscs and other aquatic invertebrates, prepared or preserved, n.e.s.
S-03412:	Salmonidae, fresh or chilled (excluding livers and roes)
S-03421:	Salmonidae, frozen (excluding livers and roes)
S-03711:	Salmon, whole or in pieces, but not minced

Data source: UN Comtrade (2010)

3.2 Major fish trading countries

The top ten exporting and importing counties of fish (SITC-code 03) are listed in Table 2. Of the top ten exporters the three Asian countries China, Thailand, and

Viet Nam together account for 22.5 percent of total world fish exports in the years from 2005 to 2008 while the European countries Norway, Spain, Denmark, and the Netherlands account for nearly 18 percent, and the North and South American countries USA, Canada, and Chile jointly contribute nearly 14 percent. World imports of fish are dominated by the USA, Japan, and member countries of the EU, which is by far the largest market for imported fish (FAO 2010).

Table 2: Top ten importing and exporting countries of fish (SITC-code 03) by average trade value for the years 2005 to 2008, [bn. US$]

Rank	Country	Average import value [bn. US$]	Share of total world imports [%]	Country	Average export value [bn. US$]	Share of total world exports [%]
1	USA	27.9	15.7	China	17.9	11.2
2	Japan	27.2	15.4	Norway	11.5	7.2
3	Spain	13.2	7.4	Thailand	10.8	6.8
4	France	10.3	5.8	USA	8.5	5.3
5	Italy	9.8	5.5	Canada	7.3	4.5
6	Germany	7.6	4.3	Viet Nam	7.2	4.5
7	United Kingdom	7.4	4.1	Spain	6.1	3.8
8	China	6.6	3.7	Chile	6.0	3.8
9	Rep. of Korea	5.4	3.0	Denmark	5.9	3.7
10	Sweden	4.4	2.5	Netherlands	5.1	3.2
	Subtotal	119.6	67.5	Subtotal	86.2	53.8
	World total	177.2		World total	160.4	

Data source: UN Comtrade (2010)

Table 3: Top ten importing and exporting countries of fresh, chilled, or frozen fish (SITC-code 034) by average trade value for the years 2005 to 2008, [bn. US $]

Rank	Country	Average import value [bn. US$]	Share of total world imports [%]	Country	Average export value [bn. US$]	Share of total world exports [%]
1	Japan	6.9	15.4	Norway	4.7	12.5
2	USA	5.2	11.6	China	3.5	9.3
3	Spain	2.8	6.2	USA	2.8	7.6
4	China	2.6	5.8	Chile	2.4	6.5
5	France	2.4	5.5	Spain	1.5	4.1
6	Germany	2.2	4.9	Netherlands	1.5	4.0
7	United Kingdom	2.0	4.4	Denmark	1.4	3.8
8	Italy	1.7	3.9	Viet Nam	1.3	3.4
9	Rep. of Korea	1.6	3.5	Canada	1.2	3.3
10	Thailand	1.5	3.4	Iceland	1.2	3.1
	Subtotal	28.7	64.7	Subtotal	21.5	57.6
	World total	44.4		World total	37.3	

Data source: UN Comtrade (2010)

The top ten exporting and importing countries of fresh, chilled, or frozen fish (SITC-code 034) by average trade value between 2005 and 2008 are listed in Table 3. The main importing countries are Japan and the USA, while the main exporting countries are Norway and China. Compared to fish (SITC-code 03) no big difference in the concentration of exports and imports exists.

Table 4 presents the top ten exporting and importing countries of salmonidae (SITC-codes 03412, 03421 and 03711) between 2005 and 2008. Norway, by far the largest producer of salmon, is also the largest exporter of salmonidae. As aquaculture production and capture of salmonidae in Sweden, Denmark, and Germany is relatively low, the remarkable exports may be explained by trade with primary imported and then processed and exported salmonidae.

The existence of Mauritania in the list of the largest exporting countries is surprising. Neither FAO´s capture and aquaculture production statistics nor UN Comtrade´s import statistics indicate that Mauritania controls sufficient quantities of salmonidae to be the seventh largest exporting country of the world for this commodity.

Demand for salmonidae is highest in Japan followed by Sweden, USA, France and Russia. The top ten importing countries import three-quarters of total world imports of salmonidae.

Table 4: Top ten importing and exporting countries of salmonidae (SITC-codes 03412, 03421 and 03711) by average trade value for the years 2005 to 2008, [bn. US$]

Rank	Country	Average import value [bn. US$]	Share of total world imports [%]	Country	Average export value [bn. US$]	Share of total world exports [%]
1	Japan	2.0	15.0	Norway	5.1	35.9
2	Sweden	1.9	14.2	Sweden	1.6	11.4
3	USA	1.1	8.6	Chile	1.6	11.1
4	France	1.1	8.2	USA	1.3	8.9
5	Russian Federation	0.8	6.1	Canada	1.0	7.2
6	Germany	0.7	4.9	United Kingdom	0.5	3.7
7	China	0.6	4.5	Denmark	0.5	3.5
8	Denmark	0.6	4.5	Mauritania	0.3	2.2
9	United Kingdom	0.6	4.2	Germany	0.3	2.1
10	Poland	0.5	4.0	Japan	0.3	1.8
	Subtotal	9.8	74.1	Subtotal	12	87.9
	World total	13.3		World total	14.2	

Data source: UN Comtrade (2010)

The composition of the top ten exporting and importing countries is, with a few exceptions, the same for fish (SITC-code 03), fresh, chilled or frozen fish (SITC-code 034) and salmonidae (SITC-codes 03412, 03421, 03711). The analysis in the following section focuses mainly on the top ten exporting and importing countries of fish, fresh, frozen or chilled fish and fresh or chilled salmon (SITC-codes 03, 034, and 03412).

4 Methods and results

In this section the accuracy of fish trade data is analyzed. In a first step, Benford's law is applied to selected trade data sets to test if the first significant digit of the recorded numbers follows a special statistical distribution, viz. Benford's law. Benford's law is used to detect fraud in e.g. tax declarations and accounting data. When this law is applied to trade data, countries could be identified whose trade data may have been fraudulently changed or falsified and therefore the data should be inspected accurately und used with care only.

In a second step, discrepancies in trade data are scrutinized. The fact, that one transaction, the movement of goods from country A to B, is recorded twice (as export in country A and as import in country B), offers a possibility to compare and judge trade statistics. The results provide information whose countries under- or overstate their ex- and/or imports.

4.1 Benford's law applied to trade data

4.1.1 Benford's law

4.1.1.1 Benford's law

Benford's law, also known as the first-digit-law, was first published by Simon Newcomb in 1881 but then fell in oblivion (Hill 1995). In 1938 Frank Benford rediscovered the first-digit-law. Benford's law defines the distribution of first significant digits in data sets (Benford 1938, Hill 1995). Intuitively one may think that the first digits of numbers are uniformly distributed (Varian 1972, Günnel and Tödter 2007), but Newcomb and Benford detected that lower valued digits appear more often than digits of higher values and they postulated the following formula for the probability P of the nonzero first significant digit i (Hill 1995):

$$P(i) = \log_{10}(1 + {}^{1}/_{i}), \qquad i \in \{1, 2, ..., 9\} \tag{1}$$

According to (1) the probability for "1" as leading digit is 30.1 percent while the probability for "9" as the first significant digit is only 4.6 percent. Benford (1938) verified his law by checking the distribution of 20 data sets with a total of 20,299 observations. Further data sets were tested by several authors and most empirical data sets obey this law (Hill 1995, Lolbert 2008).

In general, the first significant digit law does not hold for assigned numbers, like purchase numbers, or psychological influenced numbers, like supermarket prices (Hill 1995, Durtschi et al. 2004, Nigrini and Mittermaier 1997).

The number of scientific publications using Benford's law has increased considerably in the last decade. Berger et al. (2011) recorded 621 publications dealing with Benford's law between 1881 and 2010 of which more than half (358) were published as recently as between 2000 and 2010.

Already in 1972 Varian suggested to use Benford's law to check data sets for irregularities. In the last two decades Benford's law has been applied in various areas. It has been used as a method to detect fraud or data manipulation in a variety of contexts, such as accountancy and tax fraud (Nigrini 1996, 1999b, Durtschi et al. 2004, Günnel and Tödter 2007, Morrow 2010). Benford's law has also been applied to check predictions from mathematical models for plausibility (Ley 1996, Tödter 2007) and to investigate first and second digits of published statistical results (Diekmann 2007) or to verify economic research outputs like regression coefficients, standard errors and forecasts of GDP growth and inflation in Germany (Günnel and Tödter 2007, Tödter 2009). De Ceuster et al. (1998) and Giles (2007) used Benford's law to test for psychological barriers in stock markets and ebay auctions. The reliability of survey data was checked by Judge and Schechter (2009) and Schräpler (2010).

4.1.1.2 Properties and requirements of Benford's law

Benford's law satisfies two important properties: base and scale invariance (Hill 1995). While the independence of the base is unimportant for trade data, scale invariance is. The theorem of scale invariance (Pinkham 1961) states that if a data set, which follows Benford's law, is multiplied with a constant, the result is also Benford distributed. This is a desirable property when Benford's law is applied to trade data. Probably not all trade data are originally recorded in US$, thus the trade values have to be converted before they are published by UN Comtrade. The property of scale invariance thus assures that the distribution of the first digits is not biased by converting trade values to a common currency.

Schräpler (2010) and Günnel and Tödter (2007) summarize some requirements which data have to meet so that Benford's law applies. First, data should not contain a built-in maximum or minimum value, e.g. withdrawals from an automatic teller machine or the body height of persons, because the frequency of these values will occur more often in the digit analysis and will cause biased results (Nigrini 199a, Schräpler 2010). Assigned numbers, e.g. order numbers, may also lead to biased results (Nigrini 1999a). Data should not emanate from statistical procedures, like calculated means or variances, since they themselves follow certain distributions (Mochty 2002, Günnel and Tödter 2007, Schräpler 2010). Schräpler (2010) adds that the larger the sample size the better the fit to Benford's distribution should be, but this may also be fostered if a chi-square-test is applied to a large sample (see following section and Judge and Schechter 2009).

4.1.1.3 Testing for Benford's law

Whether data follow Benford's law or not can easily be tested statistically, by comparing the distribution of a data sample with a reference distribution function (Benford's law in this case). The Kolmogorov-Smirnov (K-S) test and the Chi-square (χ^2) test are the most frequently tests applied for this purpose (Morrow 2010). Whether a data set conforms with Benford's law may also be inspected with the Kuiper-test (Kuiper 1959), which is a modification of the K-S test (Giles 2007) or with the m-test (Leemis et al. 2000, Judge and Schechter 2009, Morrow 2010).

The null hypothesis of the Chi-square test, and also of the modified K-S test and the m-test, states that the frequency distribution of certain events observed in a sample is consistent with a particular theoretical distribution (F_0). The null hypothesis (H_0) and alternative hypothesis (H_1) are formulated as follows:

$$H_0: F = F_0 \quad vs. \quad H_1: F \neq F_0 \tag{2}$$

In the case of testing for Benford's law the null hypothesis states that the data follow Benford's law while the alternative Hypothesis states the opposite. The

Chi-square test statistic (χ^2) is calculated by the differences between the empirical relative frequencies (e_i) and theoretical relative frequencies (b_i) of Benford's law which are then squared and divided by the expected theoretical relative frequencies and finally multiplied with the sample size n.

$$\chi^2 = n \sum_{i=1}^{k} \frac{(e_i - b_i)^2}{b_i}, \tag{3}$$

The use of the *Chi-square test* for testing Benford's law is criticized in the literature, because the test value directly depends on the sample size n. For large samples even quite small deviations will be statistically significant and thus the Chi-square test will be more likely to reject that the data is distributed according to Benford's law for larger samples than for smaller samples (Schräpler 2010). For moderately small sample sizes the Chi-square test is often unsuitable due to its low power (Morrow 2010). Hungerbühler (2007) states the rule of thumb that, when testing for the first significant digit, the sample size n should be larger or equal to 88. The sensitivity of the chi-square test to the sample size can be problematic so that other tests with "correction factors", like the Kuiper-test, have been developed, whose fairly accurate test statistics are insensitive to sample size (Judge and Schechter 2009, Morrow 2010).

The *Kuiper–test* identifies the largest deviation between the empirical $(F_e(x))$ and the theoretical $(F_b(x))$ cumulative distribution function (*cdf*) which is then weighted with the (inverse) square root of the sample size and a correction factor. The formula for the modified Kuiper-test is:

$$V_n^* = V_n(n^{1/2} + 0.155 + 0.24n^{-1/2}), \tag{4}$$

where $\qquad V_n = max_x[F_e(x) - F_b(x)] + max_x[F_b(x) - F_e(x)].$ (5)

In the calculation of Kuiper's V_n statistic $F_e(x)$ is the empirical observed *cdf* while $F_b(x)$ is Benford's *cdf*. The calculated value of V_n is a measure of the agreement between the empirically observed distribution and Benford's law (Arsham 1988). A large value of V_n tends to reject the null hypothesis, which is that the empirical observed sample is Benford distributed.

The *m-test* is the maximum of the absolute differences between the empirical observed and the theoretical probability distribution function (*pdf*) weighted with the square root of the sample size n. The m-test is calculated by

$$m^* = n^{1/2} \times max_{i=1,2,\ldots,9}\{|b_i - e_i|\}, \tag{6}$$

where b_i and e_i are defined as in the Chi-square test (3).

Asymptotic valid critical values were calculated for the m-test and for the Kuiper test by Morrow (2010) and are presented in Table 5. The critical values are used in the tests for Benford's law in this paper.

Table 5: Benford specific test values

Test statistic	Test level			Source
	a=0.1	a=0.05	a=0.01	
Chi-square test (χ^2)	13.36	15.51	20.09	Judge and Schechter (2009)
Kuiper test (V_n^*)	1.19	1.32	1.58	Morrow (2010)
m test (m^*)	0.85	0.97	1.21	Morrow (2010)

4.1.2 Results of testing for Benford's law

Benford's law is not only applicable to the first digit. Also second and higher order analyses have been conducted (Diekmann 2007). However, the following results present only an analysis of the first digit, as it is recommended by Günnel and Tödter (2007) for checks of data manipulation. In this study the import and export values of the top ten exporting and importing countries with all trading partner countries have been analyzed for the period from 1992 to 2008.

4.1.2.1 Fish (SITC-code 03)

Table 6 presents the results of the first digit analysis of import and export data of fish. Overall, the export data for fish do not conform with Benford's law and for some countries the reported import data deviate significantly from Benford. The null hypothesis is rejected in these cases. Fish import and export data from Germany and the Netherlands do not conform with Benford's law. The null hypothesis is rejected at least by one test, indicating a significant deviation from Benford's law for the export data of Canada, Chile, Denmark, Italy, Norway, Republic of Korea, Sweden, Thailand and the USA, and for the import data of the United Kingdom.

4.1.2.2 Fresh, chilled or frozen fish (SITC-code 034)

Table 7 presents the results of testing trade data of fresh, chilled, or frozen fish (SITC-code 034) for their compliance with Benford's law. The null hypothesis is rejected for the import and export data series for all countries. Contrary, the null hypothesis is not rejected for the selected top-ten exporting and importing countries. Significant deviations could be detected for the import and export data for Canada and the Netherlands. Export data from Chile, France, Iceland, Japan and the USA do not conform with Benford's law. The null hypothesis is also rejected for import data from Germany, Thailand and the United Kingdom. For the remaining seven countries the null hypothesis is not rejected and no significant deviation could be discovered.

4.1.2.3 Fresh or chilled salmonidae (SITC-code 03412)

The results of Benford's test for trade data of fresh and chilled salmonidae are given in Table 8. The null hypothesis is not rejected for import and export data of all countries. The Kuiper test indicates that the export data for the top ten exporters and importers deviate significantly from Benford's law while the m-test reveals that the import data do not conform to Benford's law, too.

For a sample size of $n \geq 88$ import data from Germany, Japan, Norway and Poland do not conform with Benford's law, while for export data this is only the case for Norway and the USA.

Table 6: Testing trade data of top ten ex- and importer of fish (SITC- code 03) for Benford's Law, 1992-2008[1]

Country	Trade-flow	n	χ^2	V_n*	m*	first digit								
						1	2	3	4	5	6	7	8	9
						expected frequency [%]								
						30.1	17.6	12.5	9.7	7.9	6.7	5.8	5.1	4.6
						observed frequency [%]								
all	Ex	91349	***	***	***	30.6	17.5	12.3	9.4	8.0	6.7	5.9	5.1	4.6
countries	Im	103656		*		30.1	17.5	12.4	9.6	7.9	6.9	5.8	5.2	4.7
Top-Ten	Ex	30798	**	***	**	30.8	17.6	12.3	9.2	7.8	6.4	5.9	5.2	4.8
Exp. & Imp.	Im	28402				30.1	17.8	12.3	9.8	7.8	6.5	5.8	5.0	4.8
CAN	Ex	1668	**			29.5	17.8	11.9	9.7	7.6	5.3	7.2	6.3	4.7
	Im	2051				29.7	17.9	12.6	9.6	7.2	6.1	7.2	5.2	4.6
CHL	Ex	1664			*	32.3	17.4	12.7	7.9	8.2	6.4	5.9	4.6	4.6
	Im	528				30.7	15.9	13.3	7.8	8.1	6.4	7.4	5.5	4.9
CHN	Ex	1576				31.9	17.3	12.0	9.8	6.8	6.7	5.8	5.3	4.4
	Im	1392				29.5	18.6	11.9	9.1	7.9	6.8	6.3	4.7	5.2
DEU	Ex	1929		*	*	32.1	18.5	11.7	9.5	7.5	6.2	5.2	5.2	4.1
	Im	2252		**		28.6	18.4	14.1	9.9	8.5	6.2	5.2	4.6	4.5
DNK	Ex	1778	**	***	**	32.2	16.4	10.0	9.1	9.1	7.2	5.5	5.5	5.2
	Im	1387				29.4	17.9	13.1	9.5	8.2	6.3	4.9	6.6	4.2
ESP	Ex	1955				29.0	18.1	13.3	10.9	7.2	5.6	6.3	4.6	5.0
	Im	2065				29.7	19.3	12.4	10.1	7.2	6.5	5.4	5.4	4.0
FRA	Ex	2115				30.3	18.3	12.6	8.8	7.7	5.5	6.1	5.9	4.9
	Im	2478	**	**	*	28.4	17.7	11.3	10.3	7.9	8.2	6.2	5.3	4.7
GBR	Ex	2134				31.3	16.2	12.6	8.6	9.3	6.9	5.2	5.0	5.0
	Im	1958		*		31.8	17.1	11.0	9.3	7.8	6.5	6.0	5.3	5.4
ITA	Ex	1716	*	***	***	34.0	16.3	11.3	9.6	7.4	6.7	5.5	4.4	4.7
	Im	1849				30.2	17.6	12.7	10.3	7.7	6.5	5.5	4.4	4.9
JPN	Ex	1514				28.3	17.6	13.5	9.5	9.0	6.7	6.0	4.8	4.6
	Im	2248				31.2	16.4	12.6	10.0	8.0	6.1	5.7	5.5	4.5
KOR	Ex	1400			*	27.6	19.0	12.1	9.8	8.4	7.6	6.1	4.9	4.6
	Im	1457				30.1	18.4	11.7	10.4	7.2	6.0	5.2	5.5	5.4
NLD	Ex	2175	**	**	**	29.6	20.1	12.4	8.9	7.2	5.7	5.5	5.8	4.9
	Im	1784	***	**		30.8	15.8	12.3	9.1	7.1	7.1	5.8	5.5	6.5
NOR	Ex	2207	***	***	**	31.6	20.2	12.9	9.0	7.1	5.6	5.5	3.6	4.6
	Im	1110				29.6	19.1	13.7	8.5	8.7	6.4	5.6	4.0	4.6
SWE	Ex	1010	***	***	***	30.7	19.4	11.2	5.7	8.1	6.9	6.0	6.4	5.5
	Im	1275				31.0	18.5	11.2	11.1	7.8	5.8	6.1	4.1	4.3
THA	Ex	2718		*	*	31.9	15.9	12.9	9.1	7.5	6.6	6.4	5.2	4.5
	Im	1603				31.1	19.1	11.3	9.2	9.0	6.1	5.6	4.1	4.5
USA	Ex	2202	*	**	**	29.8	15.2	12.2	9.9	8.5	7.8	6.3	5.6	4.9
	Im	2447				31.6	16.6	12.3	10.4	7.0	6.0	5.8	4.9	5.4
VNM	Ex	1037				30.0	15.8	12.7	10.8	6.1	7.2	6.4	6.1	4.9
	Im	518				27.8	18.0	13.7	11.8	7.7	7.5	6.2	3.5	3.9

* indicates 90%, ** indicates 95% and *** indicates 99% significantly different from Benford.

---

[1] Table A 5 in the appendix presents a list of country abbreviations.

Table 7: Testing trade data of top ten ex- and importer of fresh, chilled or frozen fish (SITC- code 034) for Benford's Law, 1992-2008[2]

Country	Trade-flow	n	χ^2	V_n*	m*	1	2	3	4	5	6	7	8	9
						first digit								
						30.1	*17.6*	*12.5*	*9.7*	*7.9*	*6.7*	*5.8*	*5.1*	*4.6*
						expected frequency [%] / observed frequency [%]								
all	Ex	65684	***	***	*	30.1	17.7	12.1	9.4	8.2	6.8	6.0	5.2	4.6
countries	Im	74745	**	***	*	30.2	17.3	12.4	9.5	8.1	6.8	5.8	5.2	4.8
Top-Ten	Ex	22053				30.1	18.0	12.1	9.4	8.1	6.6	5.9	5.2	4.6
Exp. & Imp.	Im	23125				30.4	17.3	12.3	9.7	8.0	6.7	5.8	5.2	4.6
CAN	Ex	1027	*	***	***	34.3	17.0	10.2	8.7	7.9	5.6	5.6	5.7	5.2
	Im	1710		**		28.6	18.4	14.1	10.6	8.2	6.6	5.1	4.6	3.9
CHL	Ex	1178		**	***	31.6	14.0	12.1	10.2	7.8	7.3	6.0	6.1	4.8
	Im	265				29.8	13.6	10.6	11.7	6.4	8.3	8.7	5.7	5.3
CHN	Ex	1157				30.3	17.0	11.0	9.9	9.0	7.8	5.1	6.1	4.0
	Im	1115				30.3	17.0	11.5	9.3	7.5	7.6	7.2	5.2	4.3
DEU	Ex	1344				30.1	19.4	11.6	8.9	8.7	6.6	5.8	4.8	4.1
	Im	1927	*	***		32.0	19.2	12.9	8.1	7.4	6.1	5.4	4.7	4.4
DNK	Ex	1213				28.2	18.4	12.0	9.5	7.3	7.8	7.0	5.9	4.0
	Im	1158				29.5	17.8	13.0	10.4	9.0	5.5	5.4	6.0	3.4
ESP	Ex	1523				30.7	16.2	11.9	10.7	7.0	7.2	6.1	4.7	5.6
	Im	1799				30.7	17.0	11.8	10.0	8.6	6.3	5.7	5.6	4.5
FRA	Ex	1602	***			29.6	16.0	13.4	10.2	8.6	5.0	5.4	6.6	5.2
	Im	2055				29.4	18.0	11.2	9.6	8.0	8.0	5.6	5.2	5.1
GBR	Ex	1629				31.9	18.0	11.4	10.0	8.2	6.3	5.1	4.5	4.7
	Im	1736		**	***	33.2	17.0	12.2	8.6	7.3	6.3	5.4	5.4	4.6
ISL	Ex	745	**			32.0	16.5	12.1	9.9	9.3	8.2	5.6	4.3	2.2
	Im	391				31.2	20.5	12.0	7.7	9.0	6.4	3.8	5.6	3.8
ITA	Ex	961				30.9	18.9	11.8	8.4	8.8	5.6	5.9	4.9	4.7
	Im	1618				31.8	17.4	12.7	9.0	8.0	6.6	5.7	4.6	4.3
JPN	Ex	1134		**	***	25.8	19.0	12.5	10.1	7.9	7.9	6.2	5.3	5.3
	Im	1972				29.9	15.8	12.2	11.4	8.0	6.6	6.1	5.2	4.8
KOR	Ex	948				27.9	18.3	13.0	9.3	8.8	5.0	7.1	5.4	5.5
	Im	1202				30.3	17.4	11.7	11.0	7.7	5.8	5.7	5.3	5.0
NLD	Ex	1941	**	**		30.7	19.4	11.5	7.9	8.1	6.0	5.6	5.2	5.6
	Im	1516	***	***		28.1	15.8	12.3	9.1	6.9	7.6	7.1	6.5	6.7
NOR	Ex	1613				30.6	19.3	12.7	8.8	7.4	6.2	5.8	4.9	4.2
	Im	880				32.5	17.1	12.1	7.7	7.8	6.5	5.7	5.5	5.2
THA	Ex	1420				30.8	19.6	12.7	8.2	8.2	5.8	6.6	4.2	4.0
	Im	1276		*		30.8	17.9	12.6	11.0	8.9	5.4	5.3	3.8	4.2
USA	Ex	1837	*	**	**	27.4	19.9	11.9	9.9	8.3	7.8	5.4	5.2	4.2
	Im	2098				29.7	15.8	12.5	9.4	8.6	7.6	6.3	5.2	4.7
VNM	Ex	781				30.1	16.1	15.4	9.7	7.3	6.3	6.3	5.0	3.8
	Im	407				27.0	19.4	13.3	10.3	9.1	6.4	7.1	3.4	3.9

* indicates 90%, ** indicates 95% and *** indicates 99% significantly different from Benford.

---

[2] Table A 5 in the appendix presents a list of country abbreviations.

Table 8: Testing trade data of top ten ex- and importer of fresh or chilled salmonidae (SITC- code 03412) for Benford's Law, 1992-2008[3]

Country	Trade-flow	n	χ^2	V_n*	$m*$	1	2	3	4	5	6	7	8	9
						first digit — expected frequency [%]								
						30.1	17.6	12.5	9.7	7.9	6.7	5.8	5.1	4.6
						observed frequency [%]								
all	Ex	12223				30.7	17.9	12.4	9.1	7.6	6.8	6.0	4.9	4.5
countries	Im	11329				29.8	17.3	12.3	9.9	8.1	6.9	6.0	5.1	4.6
Top-Ten	Ex	5731		*		30.9	18.0	12.8	8.6	7.6	6.6	6.0	4.8	4.7
Exp. & Imp.	Im	3185			*	30.2	18.7	10.9	9.6	7.9	6.5	6.4	4.8	4.9
CAN	Ex	258				30.2	20.5	13.6	10.1	7.8	7.0	5.8	3.5	1.6
	Im	346				26.3	19.9	11.0	11.0	8.7	6.7	7.8	4.6	4.1
CHL	Ex	380				27.4	16.3	12.9	9.2	10.3	9.0	5.3	5.8	4.0
	Im	5			*	20.0	60.0	0.0	0.0	0.0	0.0	0.0	20.0	0.0
DEU	Ex	336				33.0	17.3	10.4	9.8	7.7	6.9	7.1	6.0	1.8
	Im	398	***	***	***	37.4	19.1	13.6	9.8	6.8	4.5	2.5	3.8	2.5
DNK	Ex	542				29.3	18.6	12.7	8.9	6.3	6.1	6.5	5.9	5.7
	Im	336				31.3	18.2	11.6	8.6	6.3	8.3	6.6	2.4	6.9
ESP	Ex	252				30.6	19.4	11.1	8.7	7.9	7.1	4.0	4.4	6.8
	Im	308				27.3	17.9	13.3	11.7	6.5	7.8	6.2	4.6	4.9
FOR	Ex	252				31.0	13.9	13.5	9.5	6.4	6.4	7.5	5.2	6.8
	Im	29				37.9	10.3	6.9	10.3	6.9	3.5	6.9	6.9	10.3
FRA	Ex	504				31.6	18.1	12.3	7.3	8.3	5.0	6.0	6.6	5.0
	Im	450				31.3	20.7	10.4	8.2	7.8	5.6	5.3	5.8	4.9
GBR	Ex	761		*		32.6	18.0	12.6	7.8	6.7	5.8	5.5	5.1	5.9
	Im	268				31.7	17.9	9.0	9.3	10.8	4.1	8.2	4.9	4.1
JPN	Ex	24	**	***	***	4.2	37.5	25.0	16.7	4.2	4.2	4.2	0.0	4.2
	Im	179	*		**	29.6	17.9	5.0	11.2	8.9	8.9	7.8	3.4	7.3
NOR	Ex	1108	**		*	32.8	17.6	13.1	7.1	7.8	6.8	6.9	4.2	3.9
	Im	142	**	***		25.4	16.9	7.0	7.0	12.0	7.0	7.8	9.9	7.0
PAN	Ex	207				31.4	18.4	12.6	9.7	7.3	5.3	4.8	2.9	7.7
	Im	10				30.0	30.0	0.0	0.0	10.0	0.0	10.0	10.0	10.0
POL	Ex	120				30.0	15.8	10.0	10.0	10.0	9.2	5.8	3.3	5.8
	Im	102	**	**	***	42.2	12.8	7.8	8.8	3.9	8.8	9.8	3.9	2.0
RUS	Ex	39		**	***	51.3	15.4	7.7	12.8	5.1	0.0	0.0	5.1	2.6
	Im	123			**	21.1	17.9	16.3	11.4	8.1	5.7	8.9	4.1	6.5
SWE	Ex	365				33.7	18.6	12.1	10.4	6.0	5.8	4.9	4.1	4.4
	Im	176				30.1	16.5	10.2	6.3	11.4	5.7	6.8	6.3	6.8
USA	Ex	583	*	**	**	25.4	19.0	15.6	9.1	8.4	8.4	5.8	4.0	4.3
	Im	313				26.2	20.8	11.8	10.9	6.7	8.3	6.4	5.1	3.8

* indicates 90%, ** indicates 95% and *** indicates 99% significantly different from Benford.

---

[3] Table A 5 in the appendix presents a list of country abbreviations.

4.2 Discrepancies in trade data

Some trade statistics have the property that they are recorded twice: once by the exporting country as export and once by the importing country as import. This offers the opportunity to test the quality of bilateral trade data by comparing both values. Mirror statistics can be used to check if the value of a country's export matches the corresponding import value of the destination country (OECD 2001). In principle the records of the importing and exporting country of the trade flow of a given good should be equal. However, this is rarely the case (OECD 2001). Reasons for discrepancies in trade data are manifold. UN (2004) divides errors in registration errors and processing errors. The major causes of registration errors include the treatment of low-value transactions; failure to file the required documentation, including smuggling and other unregistered cross-border trade; errors and missing or incomplete information; and intentionally incorrect reporting to avoid tariffs or quotas. Processing errors in trade statistics involve errors in coverage, time of recording, commodity classification, valuation (cif/fob ratios), quantity measurement and partner country attribution (UN 2004).

Specific reasons for errors in fish trade are given by FAO (2010): Customs authorities have difficulties to identify species because of a lack of reliable methods and the standard classifications used are outdated, so that they do not provide opportunities to identify "new" species and products. Another problem, which seems to be very common for seafood and might also affect the quality of trade statistics, is the renaming and mislabeling of fish and fish products (Jacquet and Pauly 2008). Opportunities for the mislabeling relevant for trade statistics are factors like the species or country of origin. Jacquet and Pauly (2008) find that mislabeling is most often done by distributors and the final seafood retailer for the sake of increased profits. If the distributor imports or exports fish it might be that incorrect commodity codes and/or values are recorded.

Moreover, different data-collection procedures may also lead to asymmetries in trade data. If the import data are derived from custom records while the export data are based on sampling techniques, differences may occur and hence it could be argued that import documentation is more complete than export documentation (UN 2004). Also reporting errors may lead to serious discrepancies in mirror statistics (UN 2004).

4.2.1 Discrepancies in world fish trade data

In a first step trade data are compared by calculating the ratio of total world exports ($X_{i\,World}$) and imports ($M_{j\,World}$).

$$R_{XM} = \frac{\sum_{i=1}^{n} X_{i\,World}}{\sum_{j=1}^{n} M_{j\,World}} \tag{7}$$

The results, presented as box plots in Figure 3, indicate that reported exports do not equal reported imports. The ratios of fish (SITC-code 03), fresh, chilled or frozen fish (SITC-code 034) as well as crustaceans and molluscs (SITC-code 036) are below one, which means that the reported imports are higher than the corresponding exports. This makes sense because the cif-valued imports exceed the fob-valued exports. The ratios for all other products may be below or larger than one. The results indicate that the differences between import and export values may depend on the product and/or the aggregation level.

Figure 3: Box plot of the ratio of exports and imports, 1992-2008

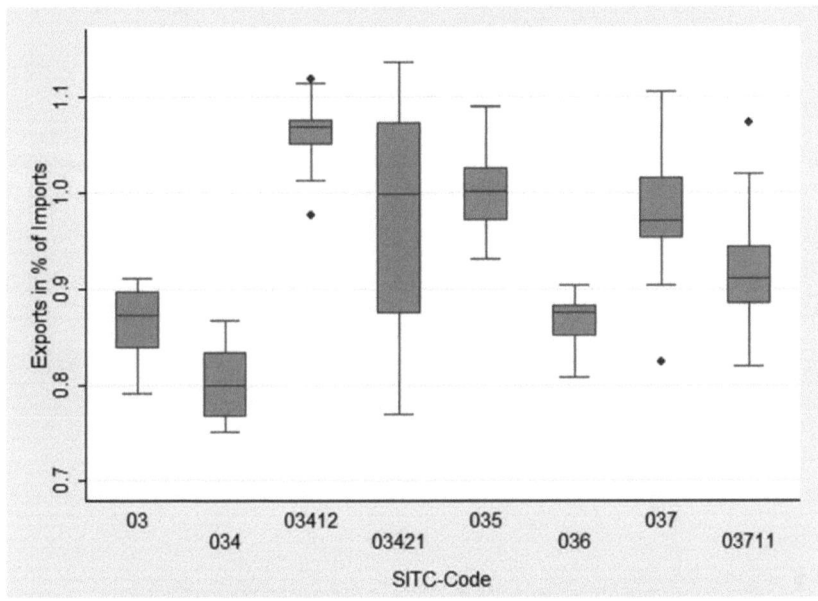

03	Fish (not marine mammals), crustaceans, molluscs and aquatic invertebrates, and preparations thereof
034:	Fish fresh (live or dead), chilled or frozen
03412:	Salmonidae, fresh or chilled (excluding livers and roes)
03421:	Salmonidae, frozen (excluding livers and roes)
035:	Fish, dried, salted or in brine; smoked fish (whether or not cooked before or during the smoking process)
036:	Crustaceans, molluscs and aquatic invertebrates, whether in shell or not, fresh (live or dead), chilled, frozen, dried, salted or in brine; crustaceans, in shell, cooked by steaming or boiling in water
037:	Fish, crustaceans, molluscs and other aquatic invertebrates, prepared or preserved, n.e.s.
03711:	Salmon, whole or in pieces, but not minced

4.2.2 Discrepancies in bilateral fish trade data

4.2.2.1 Percentage differences

Percentage differences of reported trade flows (Q) may be calculated as follows:

$$Q_1 = \frac{M_{ji}-X_{ij}}{M_{ji}} * 100 \qquad \text{for } i \neq j \tag{8}$$

$$Q_2 = \frac{X_{ji}-M_{ij}}{X_{ji}} * 100 \qquad \text{for } i \neq j \tag{9}$$

43

Based on the percentage differences (Q_1 and Q_2) Morgenstern (1950) distinguishes four cases depending on the values of Q_1 and Q_2, which are characterized in Table 9.

Table 9: Morgenstern cases

Case	Value of Q	Description
1	$Q_1 > 0$ & $Q_2 > 0$	Either country *i* overstates its exports and imports, or country *j* understates its exports and imports, or both.
2	$Q_1 > 0$ & $Q_2 < 0$	Imports are overstated in both countries or exports are understated in both countries, or both.
3	$Q_1 < 0$ & $Q_2 > 0$	Exports are overstated in both countries or imports are understated in both countries, or both.
4	$Q_1 < 0$ & $Q_2 < 0$	Either country *j* overstates its exports and imports, or country *i* understates its exports and imports, or both.

The Morgenstern approach is extended by introducing a tolerance limit to accommodate some of the discrepancies which appear to be unavoidable (see Section 4.2). If for example the tolerance limit is set to ± 10 percent and Q1 and Q2 fall in the range between 90 and 110 percent ($90 \leq Q1 \leq 110$ and $90 \leq Q2 \leq 110$) than the new created case "0" is assigned to this country pair.

4.2.2.2 Concentration indices

For each combination of countries the frequencies of the Morgenstern cases are computed and indices are calculated on the frequencies to check for the consistency and stability of the cases. Employed indices are the Herfindahl index and the MADU-0-1 index. The indices provide information on the distribution of the cases 1 to 4 for each country pairing over the years 1992 to 2008 and if trade data are constantly and systematically over- or understated by one country.

The Herfindahl index (HI) is easily computed by summing the square of the relative frequencies (s) of the four Morgenstern cases ($k = 1,..., 4$).

$$HI = \sum_{k=1}^{n} s_k^2 \tag{10}$$

If a degenerate distribution is observed (one case has the frequency of 100 percent) the Herfindahl index is one and if the cases are equally distributed the Herfindahl index reaches its minimum at 0.25 for four cases.

The mean absolute deviation from an uniform distribution (MADU) is calculated as:

$$MADU = \frac{1}{n}\sum_{k=1}^{n} \left| s_k - \frac{1}{n} \right| \tag{11}$$

In the case of a uniform distribution MADU is zero and in the case of a degenerate distribution for n=4 MADU takes a value of 0.375. To scale MADU between zero and one we divide MADU by 0.375 to obtain the MADU-0-1 index:

$$MADU\ 0-1 = MADU * \frac{8}{3} \tag{12}$$

4.2.2.3 Preparation of the data set

The calculation of the Morgenstern cases requires that export and import data are available for a country pair. When merging the import and export data sets, some observations had to be dropped because one of the two necessary values was missing. Table 10 gives an overview on the number of import and export observations and on the number of merged country pairs. It seems, the more specific the trade data get, the less the probability that the same trade flow is reported by exporter and importer.

Table 10: Number of observations in analyzed data sets

	Fish	Fresh, chilled or frozen fish	Fresh or chilled salmonidae
		SITC-code	
	03	034	03412
import data set	4,495	4,257	1,477
export data set	4,418	4,121	1,559
merged country pairs	4,226	3,626	662

4.2.3 Results of Morgenstern classification

Percentage differences of trade data were calculated and then classified into the Morgenstern classes for the commodity classifications fish (not marine mammals), crustaceans, molluscs and aquatic invertebrates, and preparations thereof (SITC-code 03), fish fresh (live or dead), chilled or frozen (SITC-code 034) and salmonidae, fresh or chilled (excluding livers and roes) (SITC-code 03412). This analysis was done for the top ten exporting and importing countries of each of the fish product categories and for the period from 1992 to 2008.

4.2.3.1 Fish (SITC-code 03)

The range of discrepancies between import and export values is large for fish. In particular, exports from France to Chile in 2006 were 2,516 times higher than the reported imports. Moreover, exports from Sweden to Thailand in 1993 accounted for only 0.02 percent of the import value reported by Thailand.

Table A 1 in the appendix shows the detailed Morgenstern classification of bilateral trade data for each country combination for a ±10 percent tolerance level.

As the cases 0 to 4 are all existent for the country combination Canada (CAN) – Viet Nam (VNM), this country pair is selected to give an example for the interpretation of the results. Common export and import data were available for ten years. In 2 cases the trade data of both countries lay within the limit of tolerance by ±10 percent (case 0). In one year Viet Nam overstated its foreign trade statistics for exports and imports or Canada understated its foreign trade in both for exports and imports (case 1). In two out of ten analyzed years, imports were overstated or exports were understated in both countries (case 2) and in one year exports were overstated or imports were understated in both countries (case 3). In the four years in which case 4 applies Viet Nam understated its foreign trade statistics or Canada overstated its foreign trade statistics for exports and imports. The relatively low levels of Herfindahl-Index and MADU-0-1 index indicate a non-concentrated distribution between the cases one to four. In the case of Canada and Viet Nam it cannot be said that one country systematically overstates or understates it trade values.

Contrary to the case of Canada and Viet Nam, the concentration indices are one for fish trade between Canada and the USA. For this country pair exports are overstated or imports are understated by both countries for the whole period from 1992 to 2008. It might be assumed that the trade statistics of fish between Canada and the USA are biased systematically. Further degenerated distributions are detected for Japan and USA, Chile and the Netherlands, Norway and Denmark, Norway and Thailand, Germany and Viet Nam, Spain and France, Italy and Viet Nam. For these country combinations it might be that e.g. one country always overstates its imports or understates its exports. This conclusion cannot be drawn, if each case (1-4) has the same probability.

Case 0 is of special interest because it indicates that the reported trade data deviate by less than ±10 percent and thus the trade data of a country pair may be considered as consistent. High frequencies for case 0 were detected for following country combinations: Italy and Spain, Japan and the Republic of Korea, France and Italy, France and the United Kingdom.

Table 11 shows that case 2 has the highest frequency in the case of fish (SITC-code 03) for a tolerance limit of zero. This indicates that imports are mainly valued higher than exports and may be a result of the different valuation of imports (cif) and exports (fob). Table 5 also indicates the sensitivity of the results if the level of tolerance is changed between 0 percent up to ±75 percent. The frequency of case 1 and 3 shrinks faster with an increasing level of tolerance as it is the case for Morgenstern classes 2 and 4. But, it is noteworthy that still more than 50 percent of the trade data deviate by more than ±50 percent from their counterpart. This indicates that the mirror statistics of many country pairs do not coincide with each other and thus the statistics seem to be of low quality.

Table 11: Frequency of cases 0 to 4 as relative frequencies [%] depending on accepted tolerance limit for fish (SITC-code 03), 1992 - 2008 (N=4,226)

accepted discrepancy [%]	case 0	case 1	case 2	case 3	case 4
0	-	23.7	37.4	15.2	23.7
± 5	1.5	23.3	37.1	14.8	23.4
± 10	5.1	22.3	35.9	14.1	22.5
± 15	9.2	21.2	34.4	13.3	21.9
± 20	14.7	19.5	32.8	12.3	20.7
± 30	28.1	15.4	27.8	10.4	18.3
± 50	49.6	9.3	19.5	7.3	14.3
± 75	66.6	4.2	12.8	4.6	11.7

Case 0:	Percentage differences Q_1 and Q_2 are within the tolerance limit.
Case 1:	Percentage differences Q_1 and Q_2 are greater than the accepted tolerance limit. Country *i* overstates or country *j* understates foreign trade statistics.
Case 2:	Percentage difference Q_1 is greater than the accepted tolerance limit and Q_2 is less than the accepted tolerance limit. Imports are overstated or exports are understated. in both countries
Case 3:	Percentage difference Q_1 is less than the accepted tolerance limit and Q_2 is greater than the accepted tolerance limit. Imports are understated or exports are overstated in both countries.
Case 4:	Percentage differences Q_1 and Q_2 are less than the accepted tolerance limit. Country *j* overstates or country *i* understates foreign trade statistics.

4.2.3.2 Fresh, chilled or frozen fish (SITC-code 034)

Extreme discrepancies could also be detected for trade with commodities of fresh, chilled or frozen fish (SITC-code 034). While the exports to the Republic of Korea from Italy in 2006 accounted for only 0.06 percent of the reported imports, the exports from Norway to Italy in 2001 were 64,011 times higher than the imports reported by Italy.

Table A 2 in the appendix shows the Morgenstern classification with a tolerance limit of ±10 percent for fish (SITC-code 034). Comparatively high frequencies of fish trade data from Italy, Republic of Korea, Japan and the Netherlands lie in between the boundaries of the ±10 percent tolerance limit (case 0). Contrary, for Chile no case "0" could be detected. Degenerate distributions of the cases 1 to 4 could be observed for nearly each country with the exception of Viet Nam, United Kingdom, Spain, Denmark, Norway and the Netherlands. In contrast, cases 1 to 4 are comparatively uniformly distributed for the Netherlands and Asian trading partners or for China and its European trading partners, to name just a few examples. For some country pairs a systematic bias could be detected while the discrepancies seem to be more or less random for other country pairs.

The distribution of cases 1 to 4 shows only small changes in comparison to fish (SITC-code 03), when the tolerance limit is set to zero: case 2 has the highest frequency with 40.5 percent, followed by case 1 and 4 with 23.3 percent each and case 4 with 12.9 percent (see Table 12). With an increasing tolerance limit relative frequencies in case 1 shrink faster than they do for case 4. More than 60 percent of the trade data of the top-ten importing and exporting countries of fish (SITC-code 034) deviate by more than ±50 percent. The data quality of mirror statistics seems to worsen, the more detailed the product classification is.

Table 12: Frequency of cases 0 to 4 as relative frequencies [%] depending on accepted tolerance limit for fresh, chilled or frozen fish (SITC-code 034), 1992 – 2008 (N=3,626)

accepted discrepancy [%]	case 0	case 1	case 2	case 3	case 4
0	-	23.3	40.5	12.9	23.3
± 5	0.6	23.1	40.4	12.7	23.1
± 10	3.0	22.5	39.4	12.5	22.6
± 15	6.9	21.6	38.0	11.5	22.0
± 20	10.7	20.5	36.6	10.9	21.4
± 30	21.0	17.4	32.8	9.4	19.4
± 50	38.8	11.8	26.2	6.9	16.4
± 75	56.7	5.5	18.8	5.0	14.0

Case 0:	Percentage differences Q_1 and Q_2 are within the tolerance limit.
Case 1:	Percentage differences Q_1 and Q_2 are greater than the accepted tolerance limit. Country i overstates or country j understates foreign trade statistics.
Case 2:	Percentage difference Q_1 is greater than the accepted tolerance limit and Q_2 is less than the accepted tolerance limit. Imports are overstated or exports are understated. in both countries
Case 3:	Percentage difference Q_1 is less than the accepted tolerance limit and Q_2 is greater than the accepted tolerance limit. Imports are understated or exports are overstated in both countries.
Case 4:	Percentage differences Q_1 and Q_2 are less than the accepted tolerance limit. Country j overstates or country i understates foreign trade statistics.

4.2.3.3 Fresh or chilled salmonidae (SITC-code 03412)

Large discrepancies in the trade data were also detected for trade with salmonidae. The exports from Poland to Denmark accounted for only 0.2 percent of the reported imports in 2008. The exports from the USA to Panama were reported 12,862 times higher than the imports in 2008.

The sample of trade data of the top ten exporting and importing countries of salmon is small compared to both other fish product categories. Table A 3 in the appendix gives a detailed overview of the Morgenstern classes. It is remarkable that 11 out of 17 trade records between USA and Canada lie within the ±10 percent tolerance limit. The Herfindahl index and MADU-0-1 index have low values for trade with salmonidae between the United Kingdom and Canada, Spain, France and the USA.

In contrast to fish (SITC-codes 03 and 034), case 3 has the highest frequency in trade with salmonidae (36.9 percent) and case 2 appears only in 13.9 percent of the cases (see Table 13). Additionally, the results are less sensitive to a change in the accepted discrepancy. Even if the tolerance limit is set to ±50 percent, nearly 80 percent of the data lie out of this range. It seems, the more specific trade data get the less consistent they are.

Table 13: Frequency of cases 0 to 4 as relative frequencies [%] depending on accepted tolerance limit for fresh or chilled salmonidae (SITC-code 03412), 1992 - 2008 (N=662)

accepted discrepancy [%]	case 0	case 1	case 2	case 3	case 4
0	-	24.6	13.9	36.9	24.6
± 5	1.5	24.3	13.9	36.0	24.3
± 10	4.7	23.7	13.6	34.1	23.9
± 15	7.9	23.0	13.3	32.8	23.1
± 20	8.9	22.7	13.3	32.2	23.0
± 30	11.9	22.2	13.1	30.1	22.7
± 50	20.9	19.9	10.9	26.6	21.8
± 75	34.6	13.1	8.5	23.1	20.7

Case 0:	Percentage differences Q_1 and Q_2 are within the tolerance limit.
Case 1:	Percentage differences Q_1 and Q_2 are greater than the accepted tolerance limit. Country i overstates or country j understates foreign trade statistics.
Case 2:	Percentage difference Q_1 is greater than the accepted tolerance limit and Q_2 is less than the accepted tolerance limit. Imports are overstated or exports are understated. in both countries
Case 3:	Percentage difference Q_1 is less than the accepted tolerance limit and Q_2 is greater than the accepted tolerance limit. Imports are understated or exports are overstated in both countries.
Case 4:	Percentage differences Q_1 and Q_2 are less than the accepted tolerance limit. Country j overstates or country i understates foreign trade statistics.

5   Discussion and summary

This study showed (i) that some trade data deviate significantly from Benford's law and (ii) that bilateral fish trade data are inaccurate.

Benford's law is primarily based on empirical observations. The number of scientific applications of Benford's law increased significantly in the last decade. However, it might be put into question why the first digits of trade data should follow the first digit law. Contrary, it is astonishing how many data follow Benford's law.

For some countries the null hypotheses was rejected. Here it might be assumed that irregularities occurred somewhere in the process between collecting and publishing the data. In this study the Chi-square test, the modified K-S test, and the m-test were applied, to account for the fact that the test value of the Chi-square test directly depends on the sample size.

In the case of fish (SITC-code 03) the null hypothesis was rejected for the export for all countries. This was also the case when only the data of the the top ten importing and exporting countries were analyzed. In the case of fresh, chilled or frozen fish (SITC-code 034) the trade data of all countries do not follow Benford's law. The null hypothesis was not rejected for the trade data of the top ten exporting and importing countries. The first digits of export and import data of salmonidae (SITC-code 03412) of all countries follow Benford's law and only one statistical test rejects the null-hypothesis for the top ten exporting and importing countries.

Results based on the trade data by country suggest that the export data for fish (SITC-code 03) are more likely to deviate from Benfords law. For fresh, chilled or frozen fish and salmonidae (SITC-codes 034 and 03412) no such clear conclusion can be drawn. For the three fish product categories the import data of Germany and the export data of the USA do not conform to Benford's law. For the export and import data of Spain the null hypotheses are not rejected for all three fish products. Nevertheless, the rejection of the null hypothesis for one country does not automatically mean that this country falsifies its trade statistics and vice versa.

Not only fish trade data deviate from Benford's law. The first digits of export data of 20 agricultural products have been analyzed for their compliance with Benford's law by Guettler et al. (2012). They find significant deviations from Benford's law in some years and for several exporting countries. It can be concluded that many trade data of agricultural products, including fish, deviate from Benford's law. An analysis on the quality of wine trade data by Thiemann et al. (2011) revealed large discrepancies between export and import data in the period from 2000 to 2008. They also checked the consistency among trade data from alternative sources and find partially large deviations between them. This indicates that the quality of trade data is not only a problem for fish trade data from UN Comtrade, but also for other commodities and alternative sources of trade data.

The comparison of bilateral trade data revealed large discrepancies between import and export statistics between country pairs. Percentage differences between import and export data of a country pair were calculated and classified into four cases suggested by Morgenstern (1950). A tolerance limit was introduced to accommodate some of the discrepancies which appear to be unavoidable. Some country pairs are identified whose trade data seem to be biased systematically. The discrepancies between export and import statistics for most of the country pairs seem to be more or less random. The sensitivity analysis with changing tolerance limits revealed that for all three fish products more than 50 percent of the observed trade values deviate by more than ±50 percent. The data quality of mirror statistics seems to worsen, the more detailed the product classification is. However, no systematic pattern in the discrepancies is detected.

A lot of reasons exist, why differences in mirror statistics can appear. Neither Benford's law nor the classification of Morgenstern (1950) indicates why trade data are biased and inaccurate. Further research would be needed to identify the reasons for each country pair. It could also be useful to separate the results of Benford's law for each year to get more information when and how often the trade data of a country deviate from Benford's law. Moreover, the results indicate that further quality programmes of the statistical offices are necessary to improve the trade statistics of fish and probably of other goods, too. The large discrepancies between import and export data of country pairs make it reasonable to scrutinize bilateral trade data before using them for further analyses or deducing political decisions. This also implies that the results of the next study on networks of international fish trade (Chapter 4) have to be interpreted with care.

6 References

Arsham, H. (1988): Kuiper's P-value as a measuring tool and a decision procedure for the goodness-of-fit test. Journal of Applied Statistics 15 (2): 131-135.

Benford, F. (1938): The law of anomalous numbers. Proceedings of the American Philosophical Society 78: 551–572.

Bergdahl, M., Ehling, M., Elvers, E., Földesi, E., Körner, T., Kron, A., Lohauß, P., Mag, K., Morais, V., Nimmergut, A., Sæbø, H.V., Timm, U. and Zilhão, M.J. (2007): Handbook on data quality assessment methods and tools. European Commission, Eurostat, Wiesbaden. <http://epp.eurostat.ec.europa.eu/portal/page/portal/quality/documents/H ANDBOOK%20ON%20DATA%20QUALITY%20ASSESSMENT%20MET HODS%20AND%20TOOLS%20%20I.pdf>

Berger, A., Hill, T.P. and Rogers, E. (2011): Benford online bibliography <http://www.benfordonline.net/list/chronological>, <06.07.2011>.

De Ceuster, M.J.K., Dhaene, G. and Schatteman, T. (1998): On the hypothesis of psychological barriers in stock markets and Benford's law. Journal of Empirical Finance 5: 263–267.

Diekmann, A. (2007): Not the first digit! Using Benford's law to detect fraudulent scientific data. Journal of Applied Statistics 34 (3): 321-329.

Durtschi, C., Hillison, W. and Pacini, C. (2004): The effective use of Benford's law to assist in detecting fraud in accounting data. Journal of Forensic Accounting 5: 17-34.

FAO (2010): The state of world fisheries and aquaculture 2010. Food and Agriculture Organization of the United Nations, Rome.

FAO (2011): FAO Fisheries and Aquaculture Department, Statistics and Information Service, FishStatJ: Universal software for fishery statistical time series. <http://www.fao.org/fishery/statistics/software/fishstatj/en>.

Giles, D.E. (2007): Benford's law and naturally occurring prices in certain ebay auctions. Applied Economics Letters 14: 157-161.

Günnel, S. and Tödter, K.H. (2007): Does Benford's law hold in economic research and forecasting? Deutsche Bundesbank Discussion Paper Series 1: Economic Studies, No. 32/2007. <http://www.bundesbank.de/download/volkswirtschaft/dkp/2007/200732d kp.pdf>.

Guettler, S., Thiemann, F. and Mueller, R.A.E. (2012): Benfords Gesetz: Ein Qualitätstest für statistische Reihen angewendet auf Handelsdaten für Agrarprodukte. In: Clasen, M., Fröhlich, G., Bernhardt, H., Hildebrand, K. and Theuvsen, B. (eds.): Informationstechnologie für eine nachhaltige Landbewirtschaftung. Referate der 32. GIL-Jahrestagung, 29. Februar – 1. März 2012, Freising. Lecture Notes in Informatics 194. Gesellschaft für Informatik, Bonn: 111-114.

Hill, T.P. (1995): A statistical derivation of the significant-digit law. Statistical Science 10 (4): 354-363.

Hungerbühler, N. (2007): Benford's Gesetz über führende Ziffern: Wie die Mathematik Steuersündern das Fürchten lehrt. <http://www.educ.ethz.ch/unt/um/mathe/ana/benford/Benford_Fuehrende _Ziffern.pdf>.

Jacquet, J.L. and Pauly, D. (2008): Trade secrets: Renaming and mislabeling of seafood. Marine Policy 32: 309-318.

Judge, G. and Schechter, L. (2009): Detecting problems in survey data using Benford's law. Journal of Human Resources 44 (1): 1-24.

Kuiper, N.H. (1959): Alternative proof of a theorem of Birnhaum and Pyke, The Annals of Mathematical Statistics 30 (1): 251–252.

Leemis, L.M., Schmeiser, B.W. and Evans, D.L. (2000): Survival distributions satisfying Benford's law. The American Statistician 54 (3): 1-6.

Ley, E. (1996): On the peculiar distribution of the U.S. stock indexes' digits. The American Statistician 50: 311–313.

Lolbert, T. (2008): On the non-existence of a general Benford's law. Mathematical Social Sciences 55: 103-106.

Martin (2010): What's the difference? - Comparing U.S. and Chinese Trade data. Congressional Research Service. <http://fpc.state.gov/documents/organization/122443.pdf>.

Mochty, L. (2002): Die Aufdeckung von Manipulationen im Rechnungswesen: Was leistet das Benford's law? Die Wirtschaftsprüfung 14: 725-736.

Morgenstern, O. (1950): On the accuracy of economic observations. Princeton University Press, Princeton.

Morrow, J. (2010): Benford's law, families of distributions and a test basis. Draft: October 9, 2010. <http://www.johnmorrow.info/projects/benford/benfordMain.pdf>.

Newcomb, S. (1881): Note on the frequency of use of the different digits in natural numbers. American Journal of Mathematics 4 (1): 39-40.

Nigrini, M.J. (1996): A taxpayer compliance application of Benford's law. The Journal of the American Taxation Association 18: 72–91.

Nigrini, M.J (1999a): I've got your number. Journal of Accountancy 187 (5): 79-83.

Nigrini, M.J. (1999b): Adding value with digital analysis. The Internal Auditor 56: 21-23.

Nigrini, M.J. and Mittermaier, L.J. (1997): The use of Benford's law as an aid in analytical procedures. Auditing: A Journal of Practice & Theory 16 (2): 52–67.

OECD (2001): Trade in goods and services: Statistical trends and measurement challenges. OECD Statistics Brief, October 2001 No. 1.

Pinkham, R.S. (1961): On the distribution of first significant digits. The Annals of Mathematical Statistics 32 (4): 1223-1230.

Rozanski, J., Yeats, A. (1994): Foreign Trade on the (in)accuracy of economic observations: An assessment of trends in the reliability of international trade statistics. Journal of Development Economics 44: 103-130.

Schräpler, J.P. (2010): Benford's law as an instrument for fraud detection in surveys using the data of the Socio-Economic Panel (SOEP). SOEPpapers on Multidisciplinary Panel Data Research at DIW Berlin 273, February 2010, Berlin. <http://www.diw.de/documents/publikationen/73/diw_01.c.349061.de/diw_sp0273.pdf>.

Scott, P.D. and Fasli, M. (2001): Benford's law: An empirical investigation and an novel explanation. CSM Technical Report 349. <http://citeseerx.ist.psu.edu/viewdoc/download?doi=10.1.1.1.9527&rep=rep1&type=pdf>.

Thiemann, F., Guettler, S. and Mueller, R.A.E. (2011): Discrepancies in wine trade statistics: a serious cause for concern? Paper presented at: 5th annual conference of the American Association of Wine Economists, Bozen, Italy, 22. - 25. June 2011. <http://www.wine-economics.org/meetings/Bolzano2011/Submissions/Thiemann_Guettler_Mueller.pdf> and <http://www.agric-econ.uni-kiel.de/Abteilungen/II/PDFs/bozen_v2a_web.pdf>.

Tödter, K.H. (2007): Das Benford-Gesetz und die Anfangsziffern von Aktienkursen. Wirtschaftswissenschaftliches Studium 36(2): 93–97.

Tödter, K.H. (2009): Benford's law as an indicator of fraud in economics. German Economic Review 10(3): 339-351.

UN (1991): Standard International Trade Classification Revision 3. Statistical Papers Series M No. 34/Rev. 3. United Nations Publication, New York. <http://unstats.un.org/unsd/publication/SeriesM/SeriesM_34rev3E.pdf>.

UN (2004): International Merchandise Trade Statistics – Compilers Manual. United Nations Publication, New York.

UN (2008): International Merchandise Trade Statistics – Supplement to the Compilers Manual. United Nations Publication, New York.

UN (2010): United Nations International Trade Statistics Knowledgebase: Comtrade Country Code and Name.

<http://unstats.un.org/unsd/tradekb/Knowledgebase/Comtrade-Country-Code-and-Name>, <23.07.2011>.

UN Comtrade (2010): United Nations Commodity Trade Statistics Database. <http://comtrade.un.org/db/default.aspx>.

Varian, H.R. (1972): Benford's law. The American Statistician 26(3): 65.

Yeats, A.J. (1990): On the accuracy of economic observations: Do Sub-Saharan Trade Statistics mean anything? World Bank Economic Review 4(2): 135-156. <http://wber.oxfordjournals.org/cgi/reprint/4/2/135>.

7 Appendix

Table A 1: Absolute frequencies of cases 0 to 4 with a tolerance limit of ± 10 percent by country pairs and their Herfindahl and MADU-0-1 indices for fish (SITC-code 03), 1992-2008

Country-pair		cases				HI	MADU-0-1
	0	1	2	3	4		
CAN-CHL	0	0	4	0	8	0.6	0.7
CAN-CHN	0	0	3	0	14	0.7	0.8
CAN-DEU	2	0	11	0	4	0.6	0.7
CAN-DNK	0	4	12	0	1	0.6	0.6
CAN-ESP	0	2	14	0	1	0.7	0.8
CAN-FRA	0	8	9	0	0	0.5	0.7
CAN-GBR	1	5	11	0	0	0.6	0.7
CAN-ITA	0	1	16	0	0	0.9	0.9
CAN-JPN	0	2	15	0	0	0.8	0.8
CAN-KOR	0	1	14	0	2	0.7	0.8
CAN-NLD	0	1	10	4	2	0.4	0.5
CAN-NOR	0	3	14	0	0	0.7	0.8
CAN-SWE	1	4	10	0	2	0.5	0.5
CAN-THA	0	1	14	0	2	0.7	0.8
CAN-USA	0	0	0	17	0	1.0	1.0
CAN-VNM	2	1	2	1	4	0.3	0.3
N=220	6	33	159	22	40		
%	2.3	12.7	61.2	8.5	15.4		
CHL-CAN	0	8	4	0	0	0.6	0.7
CHL-CHN	0	0	0	3	11	0.7	0.7
CHL-DEU	0	0	0	15	2	0.8	0.8
CHL-DNK	0	8	9	0	0	0.5	0.7
CHL-ESP	2	1	14	0	0	0.9	0.9
CHL-FRA	1	8	4	0	0	0.6	0.7
CHL-GBR	0	6	3	0	0	0.6	0.7
CHL-ITA	1	5	5	0	1	0.4	0.5
CHL-JPN	0	12	4	0	0	0.6	0.7
CHL-KOR	2	2	7	1	2	0.4	0.4
CHL-NLD	0	0	0	0	2	1.0	1.0
CHL-NOR	0	8	2	2	2	0.4	0.4
CHL-SWE	0	4	0	0	0	1.0	1.0
CHL-THA	0	12	2	1	0	0.7	0.7
CHL-USA	0	16	1	0	0	0.9	0.9
CHL-VNM	0	0	1	4	1	0.5	0.6
N=178	6	90	56	26	21		
%	3.0	45.2	28.1	13.1	10.6		

Country-pair		cases				HI	MADU-0-1
	0	1	2	3	4		
JPN-CAN	0	0	15	0	2	0.8	0.8
JPN-CHL	0	0	4	0	12	0.6	0.7
JPN-CHN	0	0	12	2	3	0.5	0.6
JPN-DEU	0	2	12	1	2	0.5	0.6
JPN-DNK	0	13	3	1	0	0.6	0.7
JPN-ESP	2	3	4	3	5	0.3	0.1
JPN-FRA	0	1	11	0	5	0.5	0.6
JPN-GBR	0	2	15	0	0	0.8	0.8
JPN-ITA	0	0	14	0	3	0.7	0.8
JPN-KOR	8	1	8	0	0	0.8	0.9
JPN-NLD	0	2	5	1	9	0.4	0.4
JPN-NOR	0	0	15	0	2	0.8	0.8
JPN-SWE	0	1	6	0	4	0.4	0.5
JPN-THA	1	14	2	0	0	0.8	0.8
JPN-USA	0	0	17	0	0	1.0	1.0
JPN-VNM	0	8	3	0	0	0.6	0.7
N=212	11	47	146	8	47		
%	4.2	18.1	56.4	3.1	18.1		
KOR-CAN	1	1	14	0	1	0.8	0.8
KOR-CHL	2	2	7	1	2	0.4	0.4
KOR-CHN	3	0	1	4	9	0.5	0.6
KOR-DEU	0	4	7	3	3	0.3	0.2
KOR-DNK	0	8	1	6	2	0.4	0.4
KOR-ESP	0	1	11	1	4	0.5	0.5
KOR-FRA	0	5	12	0	0	0.6	0.7
KOR-GBR	1	4	12	0	0	0.6	0.7
KOR-ITA	0	1	14	0	0	0.9	0.9
KOR-JPN	6	0	10	0	1	0.8	0.9
KOR-NLD	1	1	4	8	3	0.4	0.3
KOR-NOR	0	2	7	1	7	0.4	0.4
KOR-SWE	1	3	7	0	2	0.4	0.4
KOR-THA	2	2	7	1	5	0.4	0.4
KOR-USA	1	11	5	0	0	0.6	0.7
KOR-VNM	1	5	4	0	1	0.4	0.5
N=217	19	50	123	25	40		
%	7.4	19.5	47.9	9.7	15.6		

to be continued…

Table A 1: continued

Country-pair	cases					HI	MADU-0-1	Country-pair	cases					HI	MADU-0-1
	0	1	2	3	4				0	1	2	3	4		
CHN-CAN	0	14	3	0	0	0.7	0.8	NLD-CAN	0	2	10	4	1	0.4	0.5
CHN-CHL	0	11	0	3	0	0.7	0.7	NLD-CHL	0	2	0	0	0	1.0	1.0
CHN-DEU	0	2	8	0	7	0.4	0.5	NLD-CHN	2	6	3	5	1	0.3	0.3
CHN-DNK	0	16	1	0	0	0.9	0.9	NLD-DEU	4	0	1	5	7	0.4	0.6
CHN-ESP	0	9	8	0	0	0.5	0.7	NLD-DNK	0	14	0	2	1	0.7	0.8
CHN-FRA	0	8	8	0	0	0.5	0.7	NLD-ESP	1	15	0	1	0	0.9	0.9
CHN-GBR	0	1	8	8	0	0.4	0.6	NLD-FRA	5	1	1	4	6	0.4	0.4
CHN-ITA	0	9	5	2	0	0.4	0.5	NLD-GBR	0	1	11	0	5	0.5	0.6
CHN-JPN	0	3	12	2	0	0.5	0.6	NLD-ITA	0	7	10	0	0	0.5	0.7
CHN-KOR	3	9	1	4	0	0.5	0.6	NLD-JPN	0	9	5	1	2	0.4	0.4
CHN-NLD	2	1	3	5	6	0.3	0.3	NLD-KOR	1	3	4	8	1	0.4	0.3
CHN-NOR	1	7	6	1	2	0.4	0.4	NLD-NOR	0	5	0	11	1	0.5	0.6
CHN-SWE	1	5	7	2	1	0.4	0.4	NLD-SWE	2	5	3	6	1	0.3	0.3
CHN-THA	0	12	4	1	0	0.6	0.6	NLD-THA	0	14	0	3	0	0.7	0.8
CHN-USA	0	14	3	0	0	0.7	0.8	NLD-USA	0	14	0	3	0	0.7	0.8
CHN-VNM	0	6	1	5	0	0.4	0.6	NLD-VNM	0	5	0	2	1	0.5	0.5
N=245	7	127	78	33	16			N=221	15	103	48	55	27		
%	2.7	48.7	29.9	12.6	6.1			%	6.0	41.5	19.4	22.2	10.9		
DEU-CAN	2	4	11	0	0	0.6	0.7	NOR-CAN	0	0	14	0	3	0.7	0.8
DEU-CHL	0	2	0	15	0	0.8	0.8	NOR-CHL	0	2	2	2	8	0.4	0.4
DEU-CHN	0	7	8	0	2	0.4	0.5	NOR-CHN	1	2	6	1	7	0.4	0.4
DEU-DNK	1	12	0	4	0	0.6	0.7	NOR-DEU	0	3	14	0	0	0.7	0.8
DEU-ESP	0	0	1	16	0	0.9	0.9	NOR-DNK	1	0	0	16	0	1.0	1.0
DEU-FRA	0	0	0	16	1	0.9	0.9	NOR-ESP	0	1	1	0	15	0.8	0.8
DEU-GBR	0	8	0	9	0	0.5	0.7	NOR-FRA	1	2	1	3	10	0.4	0.5
DEU-ITA	1	15	0	1	0	0.9	0.9	NOR-GBR	0	0	2	0	15	0.8	0.8
DEU-JPN	0	2	12	1	2	0.5	0.6	NOR-ITA	0	1	0	4	12	0.6	0.6
DEU-KOR	0	3	7	3	4	0.3	0.2	NOR-JPN	0	2	15	0	0	0.8	0.8
DEU-NLD	4	7	1	5	0	0.4	0.6	NOR-KOR	0	7	7	1	2	0.4	0.4
DEU-NOR	0	0	14	0	3	0.7	0.8	NOR-NLD	0	1	0	11	5	0.5	0.6
DEU-SWE	1	12	1	3	0	0.6	0.7	NOR-SWE	1	3	12	0	1	0.6	0.7
DEU-THA	0	2	13	0	2	0.6	0.7	NOR-THA	0	0	17	0	0	1.0	1.0
DEU-USA	1	3	3	2	8	0.3	0.3	NOR-USA	1	0	13	0	3	0.7	0.8
DEU-VNM	0	8	0	0	0	1.0	1.0	NOR-VNM	0	0	3	2	7	0.4	0.4
N=241	10	85	71	75	22			N=176	5	24	107	40	88		
%	3.8	32.3	27.0	28.5	8.4			%	1.9	9.1	40.5	15.2	33.3		

to be continued…

Table A 1: continued

Country-pair	cases 0	1	2	3	4	HI	MADU-0-1	Country-pair	cases 0	1	2	3	4	HI	MADU-0-1
DNK-CAN	0	1	12	0	4	0.6	0.6	SWE-CAN	1	2	10	0	4	0.5	0.5
DNK-CHL	0	0	9	0	8	0.5	0.7	SWE-CHL	0	0	0	0	4	1.0	1.0
DNK-CHN	0	0	1	0	16	0.9	0.9	SWE-CHN	1	1	7	2	5	0.4	0.4
DNK-DEU	1	0	0	4	12	0.6	0.7	SWE-DEU	1	0	1	3	12	0.6	0.7
DNK-ESP	0	9	6	2	0	0.4	0.5	SWE-DNK	3	8	2	3	1	0.4	0.4
DNK-FRA	0	0	0	12	5	0.6	0.7	SWE-ESP	0	1	1	10	5	0.4	0.5
DNK-GBR	2	1	8	2	4	0.4	0.4	SWE-FRA	0	0	3	9	5	0.4	0.4
DNK-ITA	0	6	11	0	0	0.5	0.7	SWE-GBR	0	3	0	11	3	0.5	0.5
DNK-JPN	0	0	3	1	13	0.6	0.7	SWE-ITA	2	5	3	4	3	0.3	0.1
DNK-KOR	0	2	1	6	8	0.4	0.4	SWE-JPN	0	4	6	0	1	0.4	0.5
DNK-NLD	0	1	0	2	14	0.7	0.8	SWE-KOR	1	2	7	0	3	0.4	0.4
DNK-NOR	1	0	0	16	0	1.0	1.0	SWE-NLD	2	1	3	6	5	0.3	0.3
DNK-SWE	3	1	2	3	8	0.4	0.4	SWE-NOR	1	1	12	0	3	0.6	0.7
DNK-THA	0	1	1	0	15	0.8	0.8	SWE-THA	1	1	7	2	2	0.4	0.4
DNK-USA	1	0	5	1	10	0.5	0.6	SWE-USA	0	1	10	0	6	0.5	0.6
DNK-VNM	0	0	4	0	5	0.5	0.7	SWE-VNM	0	3	0	0	1	0.6	0.7
N=142	8	22	63	49	122			N=168	13	33	72	50	63		
%	3.0	8.3	23.9	18.6	46.2			%	5.6	14.3	31.2	21.6	27.3		
ESP-CAN	0	1	14	0	2	0.7	0.8	THA-CAN	0	2	14	0	1	0.7	0.8
ESP-CHL	2	0	14	0	1	0.9	0.9	THA-CHL	0	0	2	1	12	0.7	0.7
ESP-CHN	0	0	8	0	9	0.5	0.7	THA-CHN	0	0	4	1	12	0.6	0.6
ESP-DEU	0	0	1	16	0	0.9	0.9	THA-DEU	0	2	13	0	2	0.6	0.7
ESP-DNK	0	0	6	2	9	0.4	0.5	THA-DNK	0	15	1	0	1	0.8	0.8
ESP-FRA	1	0	0	0	16	1.0	1.0	THA-ESP	1	9	7	0	0	0.5	0.7
ESP-GBR	0	0	3	0	14	0.7	0.8	THA-FRA	0	9	8	0	0	0.5	0.7
ESP-ITA	11	1	2	1	2	0.3	0.2	THA-GBR	4	1	8	4	0	0.5	0.6
ESP-JPN	2	5	4	3	3	0.3	0.1	THA-ITA	0	10	6	1	0	0.5	0.6
ESP-KOR	0	4	11	1	1	0.5	0.5	THA-JPN	1	0	2	0	14	0.8	0.8
ESP-NLD	1	0	0	1	15	0.9	0.9	THA-KOR	2	5	7	1	2	0.4	0.4
ESP-NOR	0	15	1	0	1	0.8	0.8	THA-NLD	0	0	0	3	14	0.7	0.8
ESP-SWE	0	5	1	10	1	0.4	0.5	THA-NOR	0	0	17	0	0	1.0	1.0
ESP-THA	1	0	7	0	9	0.5	0.7	THA-SWE	1	2	7	2	1	0.4	0.4
ESP-USA	2	5	8	0	2	0.4	0.5	THA-USA	1	6	9	0	1	0.5	0.6
ESP-VNM	0	2	4	0	3	0.4	0.4	THA-VNM	0	7	0	5	0	0.5	0.7
N=176	20	38	84	34	88			N=201	10	68	105	18	60		
%	7.6	14.4	31.8	12.9	33.3			%	3.8	26.1	40.2	6.9	23.0		

to be continued…

Table A 1: continued

Country-pair	cases					HI	MADU-0-1	Country-pair	cases					HI	MADU-0-1
	0	1	2	3	4				0	1	2	3	4		
FRA-CAN	0	0	9	0	8	0.5	0.7	USA-CAN	0	0	0	17	0	1.0	1.0
FRA-CHL	1	0	4	0	8	0.6	0.7	USA-CHL	0	0	1	0	16	0.9	0.9
FRA-CHN	0	0	8	0	8	0.5	0.7	USA-CHN	0	0	3	0	14	0.7	0.8
FRA-DEU	0	1	0	16	0	0.9	0.9	USA-DEU	1	8	3	2	3	0.3	0.3
FRA-DNK	0	5	0	12	0	0.6	0.7	USA-DNK	1	10	5	1	0	0.5	0.6
FRA-ESP	1	16	0	0	0	1.0	1.0	USA-ESP	1	2	8	0	6	0.4	0.5
FRA-GBR	6	1	0	3	7	0.5	0.5	USA-FRA	0	9	8	0	0	0.5	0.7
FRA-ITA	6	6	0	1	4	0.4	0.5	USA-GBR	3	2	10	1	1	0.5	0.6
FRA-JPN	0	5	11	0	1	0.5	0.6	USA-ITA	2	0	7	0	8	0.5	0.7
FRA-KOR	0	0	12	0	5	0.6	0.7	USA-JPN	0	0	17	0	0	1.0	1.0
FRA-NLD	4	6	1	4	2	0.3	0.4	USA-KOR	2	0	4	0	11	0.6	0.7
FRA-NOR	2	9	1	3	2	0.4	0.5	USA-NLD	0	0	0	3	14	0.7	0.8
FRA-SWE	0	5	3	9	0	0.4	0.4	USA-NOR	1	3	13	0	0	0.7	0.8
FRA-THA	0	0	8	0	9	0.5	0.7	USA-SWE	0	6	10	0	1	0.5	0.6
FRA-USA	0	0	8	0	9	0.5	0.7	USA-THA	1	1	9	0	6	0.5	0.6
FRA-VNM	0	1	5	0	4	0.4	0.5	USA-VNM	0	1	3	1	7	0.4	0.4
N=193	20	55	70	48	67			N=180	12	42	101	25	87		
%	7.7	21.2	26.9	18.5	25.8			%	4.5	15.7	37.8	9.4	32.6		
GBR-CAN	1	0	11	0	5	0.6	0.7	VNM-CAN	2	4	2	1	1	0.3	0.3
GBR-CHL	0	0	3	0	6	0.6	0.7	VNM-CHL	0	1	1	4	0	0.5	0.6
GBR-CHN	0	0	8	8	1	0.4	0.6	VNM-CHN	0	0	1	5	6	0.4	0.6
GBR-DEU	0	0	0	9	8	0.5	0.7	VNM-DEU	0	0	0	0	8	1.0	1.0
GBR-DNK	2	4	8	2	1	0.4	0.4	VNM-DNK	0	5	4	0	0	0.5	0.7
GBR-ESP	0	14	3	0	0	0.7	0.8	VNM-ESP	0	3	4	0	2	0.4	0.4
GBR-FRA	6	7	0	3	1	0.5	0.5	VNM-FRA	0	4	5	0	1	0.4	0.5
GBR-ITA	5	8	0	2	2	0.5	0.6	VNM-GBR	1	0	1	6	2	0.5	0.6
GBR-JPN	0	0	15	0	2	0.8	0.8	VNM-ITA	0	0	10	0	0	1.0	1.0
GBR-KOR	1	0	12	0	4	0.6	0.7	VNM-JPN	0	0	3	0	8	0.6	0.7
GBR-NLD	0	5	11	0	1	0.5	0.6	VNM-KOR	1	1	4	0	5	0.4	0.5
GBR-NOR	0	15	2	0	0	0.8	0.8	VNM-NLD	0	1	0	2	5	0.5	0.5
GBR-SWE	0	3	0	11	3	0.5	0.5	VNM-NOR	0	7	3	2	0	0.4	0.4
GBR-THA	4	0	8	4	1	0.5	0.6	VNM-SWE	0	1	0	0	3	0.6	0.7
GBR-USA	3	1	10	1	2	0.5	0.6	VNM-THA	0	0	0	5	7	0.5	0.7
GBR-VNM	1	2	1	6	0	0.5	0.6	VNM-USA	0	7	3	1	1	0.4	0.4
N=220	23	59	92	46	37			N=105	4	34	41	26	49		
%	8.9	23.0	35.8	17.9	14.4			%	2.6	22.1	26.6	16.9	31.8		

to be continued...

Table A 1: continued

Country-pair	cases 0	1	2	3	4	HI	MADU-0-1
ITA-CAN	0	0	16	0	1	0.9	0.9
ITA-CHL	1	1	5	0	5	0.4	0.5
ITA-CHN	0	0	5	2	9	0.4	0.5
ITA-DEU	1	0	0	1	15	0.9	0.9
ITA-DNK	0	0	11	0	6	0.5	0.7
ITA-ESP	12	1	1	1	2	0.3	0.2
ITA-FRA	4	4	0	2	7	0.4	0.5
ITA-GBR	5	2	0	2	8	0.5	0.6
ITA-JPN	0	3	14	0	0	0.7	0.8
ITA-KOR	0	0	14	0	1	0.9	0.9
ITA-NLD	0	0	10	0	7	0.5	0.7
ITA-NOR	0	12	0	4	1	0.6	0.6
ITA-SWE	2	3	3	4	5	0.3	0.1
ITA-THA	0	0	6	1	10	0.5	0.6
ITA-USA	3	7	7	0	0	0.5	0.7
ITA-VNM	0	0	10	0	0	1.0	1.0
N=180	28	33	102	17	77		
%	10.9	12.8	39.7	6.6	30.0		

Table A 2: Absolute frequencies of cases 0 to 4 with a tolerance limit of ± 10 percent by country pairs and their Herfindahl and MADU-0-1 indices for fresh, chilled or frozen fish (SITC-code 034), 1992-2008

Country-pair	cases 0	1	2	3	4	HI	MADU-0-1	Country-pair	cases 0	1	2	3	4	HI	MADU-0-1
CAN-CHL	0	0	0	0	3	1.0	1.0	ISL-CAN	0	2	4	3	2	0.3	0.2
CAN-CHN	0	0	8	0	9	0.5	0.7	ISL-CHL	0	0	0	0	2	1.0	1.0
CAN-DEU	0	1	10	0	6	0.5	0.6	ISL-CHN	1	0	3	0	0	1.0	1.0
CAN-DNK	0	7	10	0	0	0.5	0.7	ISL-DEU	0	2	10	4	1	0.4	0.5
CAN-ESP	0	4	12	0	0	0.6	0.7	ISL-DNK	1	5	1	9	0	0.5	0.6
CAN-FRA	0	2	15	0	0	0.8	0.8	ISL-ESP	0	4	1	1	3	0.3	0.4
CAN-GBR	0	4	11	1	1	0.5	0.5	ISL-FRA	0	8	5	0	0	0.5	0.7
CAN-ISL	0	2	4	3	2	0.3	0.2	ISL-GBR	1	10	5	0	0	0.6	0.7
CAN-ITA	1	4	8	1	3	0.4	0.3	ISL-ITA	0	0	0	0	1	1.0	1.0
CAN-JPN	1	8	8	0	0	0.5	0.7	ISL-JPN	0	2	0	0	0	1.0	1.0
CAN-KOR	0	0	15	0	2	0.8	0.8	ISL-KOR	0	1	0	0	3	0.6	0.7
CAN-NLD	0	1	9	4	3	0.4	0.4	ISL-NLD	0	8	3	6	0	0.4	0.4
CAN-NOR	0	5	7	1	3	0.3	0.3	ISL-NOR	0	8	0	8	1	0.4	0.6
CAN-THA	1	0	14	0	2	0.8	0.8	ISL-THA	0	0	2	0	0	1.0	1.0
CAN-USA	1	0	0	16	0	1.0	1.0	ISL-USA	0	9	5	2	1	0.4	0.4
CAN-VNM	1	0	0	3	3	0.5	0.7	ISL-VNM	0	3	0	1	0	0.6	0.7
N=240	5	38	131	29	37			N=152	3	62	39	34	14		
%	2.1	15.8	54.6	12.1	15.4			%	2.0	40.8	25.7	22.4	9.2		

Country-pair	cases 0	1	2	3	4	HI	MADU-0-1	Country-pair	cases 0	1	2	3	4	HI	MADU-0-1
CHL-CAN	0	3	0	0	0	1.0	1.0	ITA-CAN	1	3	8	1	4	0.4	0.3
CHL-CHN	0	0	0	0	4	1.0	1.0	ITA-CHN	0	0	2	0	5	0.6	0.7
CHL-DEU	0	0	0	5	1	0.7	0.8	ITA-DEU	1	0	1	1	14	0.8	0.8
CHL-DNK	0	1	2	0	0	0.6	0.7	ITA-DNK	0	0	9	0	8	0.5	0.7
CHL-ESP	0	2	13	0	0	0.8	0.8	ITA-ESP	5	5	3	1	3	0.3	0.2
CHL-FRA	0	1	1	0	0	0.5	0.7	ITA-FRA	1	5	3	2	6	0.3	0.3
CHL-GBR	0	3	1	1	0	0.4	0.5	ITA-GBR	0	1	3	0	13	0.6	0.7
CHL-ISL	0	2	0	0	0	1.0	1.0	ITA-ISL	0	1	0	0	0	1.0	1.0
CHL-JPN	0	6	1	0	0	0.8	0.8	ITA-JPN	0	3	14	0	0	0.7	0.8
CHL-KOR	0	0	4	0	0	1.0	1.0	ITA-KOR	0	2	7	0	1	0.5	0.6
CHL-NOR	0	2	0	1	1	0.4	0.3	ITA-NLD	1	2	8	0	6	0.4	0.5
CHL-USA	0	14	3	0	0	0.7	0.8	ITA-NOR	0	7	0	1	0	0.8	0.8
CHL-VNM	0	0	0	1	3	0.6	0.7	ITA-THA	0	1	3	1	2	0.3	0.3
N=76	0	34	25	8	9			ITA-USA	0	1	10	1	5	0.4	0.5
%	0.0	44.7	32.9	10.5	11.8			ITA-VNM	0	2	1	1	0	0.4	0.3
								N=190	9	33	72	9	67		
								%	4.7	17.4	37.9	4.7	35.3		

to be continued...

Table A 2: continued

Country-pair	cases 0	1	2	3	4	HI	MADU-0-1	Country-pair	cases 0	1	2	3	4	HI	MADU-0-1
CHN-CAN	0	9	8	0	0	0.5	0.7	JPN-CAN	0	0	8	0	9	0.5	0.7
CHN-CHL	0	4	0	0	0	1.0	1.0	JPN-CHL	0	0	1	0	6	0.8	0.8
CHN-DEU	0	1	8	0	7	0.4	0.6	JPN-CHN	0	0	12	2	3	0.5	0.6
CHN-DNK	0	12	4	1	0	0.6	0.6	JPN-DEU	0	1	14	0	2	0.7	0.8
CHN-ESP	0	10	7	0	0	0.5	0.7	JPN-DNK	0	8	7	1	1	0.4	0.5
CHN-FRA	0	8	7	0	0	0.5	0.7	JPN-ESP	1	3	7	4	2	0.3	0.3
CHN-GBR	1	2	7	5	2	0.3	0.3	JPN-FRA	0	1	11	0	5	0.5	0.6
CHN-ISL	1	0	3	0	0	1.0	1.0	JPN-GBR	0	3	14	0	0	0.7	0.8
CHN-ITA	0	5	2	0	0	0.6	0.7	JPN-ISL	0	0	0	0	2	1.0	1.0
CHN-JPN	0	3	12	2	0	0.5	0.6	JPN-ITA	0	0	14	0	3	0.7	0.8
CHN-KOR	3	9	1	1	2	0.5	0.6	JPN-KOR	9	1	6	0	1	0.6	0.7
CHN-NLD	2	2	4	4	5	0.3	0.2	JPN-NLD	1	3	5	1	7	0.3	0.3
CHN-NOR	0	7	4	2	3	0.3	0.3	JPN-NOR	0	0	10	0	1	0.8	0.9
CHN-THA	0	8	7	2	0	0.4	0.5	JPN-THA	1	1	15	0	0	0.9	0.9
CHN-USA	0	9	8	0	0	0.5	0.7	JPN-USA	1	0	16	0	0	1.0	1.0
CHN-VNM	0	9	0	2	0	0.7	0.8	JPN-VNM	0	9	2	0	0	0.7	0.8
N=225	7	98	82	19	19			N=235	13	30	142	8	42		
%	3.1	43.6	36.4	8.4	8.4			%	5.5	12.8	60.4	3.4	17.9		
DEU-CAN	0	6	10	0	1	0.5	0.6	KOR-CAN	0	2	15	0	0	0.8	0.8
DEU-CHL	0	1	0	5	0	0.7	0.8	KOR-CHL	0	0	4	0	0	1.0	1.0
DEU-CHN	0	7	8	0	1	0.4	0.6	KOR-CHN	3	2	1	1	9	0.5	0.6
DEU-DNK	0	14	0	3	0	0.7	0.8	KOR-DEU	0	5	8	0	1	0.5	0.6
DEU-ESP	0	0	1	16	0	0.9	0.9	KOR-DNK	0	5	4	1	2	0.3	0.3
DEU-FRA	0	0	0	17	0	1.0	1.0	KOR-ESP	0	1	8	0	8	0.4	0.6
DEU-GBR	0	9	0	8	0	0.5	0.7	KOR-FRA	0	0	8	0	2	0.7	0.7
DEU-ISL	0	1	10	4	2	0.4	0.5	KOR-GBR	0	6	9	0	0	0.5	0.7
DEU-ITA	1	13	2	1	0	0.7	0.8	KOR-ISL	0	3	0	0	1	0.6	0.7
DEU-JPN	0	2	14	0	1	0.7	0.8	KOR-ITA	0	1	7	0	2	0.5	0.6
DEU-KOR	0	1	8	0	5	0.5	0.6	KOR-JPN	8	0	7	0	2	0.7	0.7
DEU-NLD	1	6	5	4	1	0.3	0.3	KOR-NLD	0	2	4	8	2	0.3	0.3
DEU-NOR	0	0	14	0	3	0.7	0.8	KOR-NOR	0	2	8	0	2	0.5	0.6
DEU-THA	0	3	8	0	5	0.4	0.4	KOR-THA	0	2	7	0	8	0.4	0.5
DEU-USA	1	5	6	2	3	0.3	0.3	KOR-USA	0	10	5	1	1	0.4	0.5
DEU-VNM	0	3	0	1	1	0.4	0.5	KOR-VNM	0	1	8	0	2	0.6	0.6
N=244	3	71	86	61	23			N=209	11	42	103	11	42		
%	1.2	29.1	35.2	25.0	9.4			%	5.3	20.1	49.3	5.3	20.1		

to be continued…

Table A 2: continued

Country-pair	cases 0	1	2	3	4	HI	MADU-0-1	Country-pair	cases 0	1	2	3	4	HI	MADU-0-1
DNK-CAN	0	0	10	0	7	0.5	0.7	NLD-CAN	0	3	9	4	1	0.4	0.4
DNK-CHL	0	0	2	0	1	0.6	0.7	NLD-CHN	2	5	4	4	2	0.3	0.2
DNK-CHN	0	0	4	1	12	0.6	0.6	NLD-DEU	1	1	5	4	6	0.3	0.3
DNK-DEU	0	0	0	3	14	0.7	0.8	NLD-DNK	1	15	0	1	0	0.9	0.9
DNK-ESP	0	11	4	2	0	0.5	0.5	NLD-ESP	1	15	0	1	0	0.9	0.9
DNK-FRA	0	0	0	10	7	0.5	0.7	NLD-FRA	2	4	2	0	9	0.4	0.5
DNK-GBR	0	1	10	2	4	0.4	0.5	NLD-GBR	0	0	8	0	9	0.5	0.7
DNK-ISL	1	0	1	9	5	0.5	0.6	NLD-ISL	0	0	3	6	8	0.4	0.4
DNK-ITA	0	8	9	0	0	0.5	0.7	NLD-ITA	1	6	8	0	2	0.4	0.5
DNK-JPN	0	1	7	1	8	0.4	0.5	NLD-JPN	1	7	5	1	3	0.3	0.3
DNK-KOR	0	2	4	1	5	0.3	0.3	NLD-KOR	0	2	4	8	2	0.3	0.3
DNK-NLD	1	0	0	1	15	0.9	0.9	NLD-NOR	0	6	1	10	0	0.5	0.6
DNK-NOR	1	3	1	12	0	0.6	0.7	NLD-THA	0	4	11	1	1	0.5	0.5
DNK-THA	0	0	14	0	3	0.7	0.8	NLD-USA	1	12	0	4	0	0.6	0.7
DNK-USA	0	1	8	0	8	0.4	0.6	NLD-VNM	0	2	2	0	3	0.3	0.3
DNK-VNM	0	0	3	0	4	0.5	0.7	N=254	10	82	62	44	46		
N=242	3	27	77	42	93			%	4.1	33.6	25.4	18.0	18.9		
%	1.2	11.2	31.8	17.4	38.4										
								NOR-CAN	0	3	7	1	5	0.3	0.3
ESP-CAN	0	0	12	0	4	0.6	0.7	NOR-CHL	0	1	0	1	2	0.4	0.3
ESP-CHL	0	0	13	0	2	0.8	0.8	NOR-CHN	0	3	4	2	7	0.3	0.3
ESP-CHN	0	0	7	0	10	0.5	0.7	NOR-DEU	0	3	14	0	0	0.7	0.8
ESP-DEU	0	0	1	16	0	0.9	0.9	NOR-DNK	0	0	1	12	4	0.6	0.6
ESP-DNK	0	0	4	2	11	0.5	0.5	NOR-ESP	0	0	0	4	10	0.6	0.7
ESP-FRA	1	0	1	0	15	0.9	0.9	NOR-FRA	1	2	2	4	8	0.3	0.3
ESP-GBR	0	0	4	0	13	0.6	0.7	NOR-GBR	0	0	2	0	15	0.8	0.8
ESP-ISL	0	3	1	1	4	0.3	0.4	NOR-ISL	0	1	0	8	8	0.4	0.6
ESP-ITA	4	3	3	1	6	0.3	0.3	NOR-ITA	0	0	0	1	7	0.8	0.8
ESP-JPN	1	2	7	4	3	0.3	0.3	NOR-JPN	0	1	10	0	0	0.8	0.9
ESP-KOR	0	8	8	0	1	0.4	0.6	NOR-KOR	0	2	8	0	2	0.5	0.6
ESP-NLD	1	0	0	1	15	0.9	0.9	NOR-NLD	0	0	1	10	6	0.5	0.6
ESP-NOR	0	10	0	4	0	0.6	0.7	NOR-THA	0	3	14	0	0	0.7	0.8
ESP-THA	1	1	5	1	9	0.4	0.5	NOR-USA	1	0	13	0	3	0.7	0.8
ESP-USA	3	2	10	0	2	0.6	0.6	NOR-VNM	0	0	2	3	6	0.4	0.4
ESP-VNM	1	4	0	0	2	0.6	0.7	N=228	2	19	78	46	83		
N=248	12	33	76	30	97			%	0.9	8.3	34.2	20.2	36.4		
%	4.8	13.3	30.6	12.1	39.1										

to be continued…

Table A 2: continued

Country-pair	cases 0	1	2	3	4	HI	MADU-0-1
FRA-CAN	0	0	15	0	2	0.8	0.8
FRA-CHL	0	0	1	0	1	0.5	0.7
FRA-CHN	0	0	7	0	8	0.5	0.7
FRA-DEU	0	0	0	17	0	1.0	1.0
FRA-DNK	0	7	0	10	0	0.5	0.7
FRA-ESP	1	15	1	0	0	0.9	0.9
FRA-GBR	3	1	1	4	8	0.4	0.5
FRA-ISL	0	0	5	0	8	0.5	0.7
FRA-ITA	1	6	3	2	5	0.3	0.3
FRA-JPN	0	5	11	0	1	0.5	0.6
FRA-KOR	0	2	8	0	0	0.7	0.7
FRA-NLD	2	9	2	0	4	0.4	0.5
FRA-NOR	1	8	2	4	2	0.3	0.3
FRA-THA	0	0	8	0	9	0.5	0.7
FRA-USA	0	0	9	0	8	0.5	0.7
FRA-VNM	0	1	4	0	3	0.4	0.5
N=235	8	54	77	37	59		
%	3.4	23.0	32.8	15.7	25.1		
GBR-CAN	1	1	10	1	4	0.5	0.5
GBR-CHL	0	0	1	1	3	0.4	0.5
GBR-CHN	1	2	7	5	2	0.3	0.3
GBR-DEU	0	0	0	8	9	0.5	0.7
GBR-DNK	0	4	10	2	1	0.5	0.7
GBR-ESP	0	13	4	0	0	0.6	0.7
GBR-FRA	3	8	1	4	1	0.4	0.5
GBR-ISL	1	0	5	0	10	0.6	0.7
GBR-ITA	0	13	3	0	1	0.6	0.7
GBR-JPN	0	0	14	0	3	0.7	0.8
GBR-KOR	0	0	9	0	6	0.5	0.7
GBR-NLD	0	9	8	0	0	0.5	0.7
GBR-NOR	0	15	2	0	0	0.8	0.8
GBR-THA	0	1	11	0	5	0.5	0.6
GBR-USA	1	1	14	0	1	0.8	0.8
GBR-VNM	0	2	0	2	1	0.4	0.4
N=245	7	69	99	23	47		
%	2.9	28.2	40.4	9.4	19.2		

Country-pair	cases 0	1	2	3	4	HI	MADU-0-1
THA-CAN	1	2	14	0	0	0.8	0.8
THA-CHN	0	0	7	2	8	0.4	0.5
THA-DEU	0	5	8	0	3	0.4	0.4
THA-DNK	0	3	14	0	0	0.7	0.8
THA-ESP	1	9	5	1	1	0.4	0.5
THA-FRA	0	9	8	0	0	0.5	0.7
THA-GBR	0	5	11	0	1	0.5	0.6
THA-ISL	0	0	2	0	0	1.0	1.0
THA-ITA	0	2	3	1	1	0.3	0.3
THA-JPN	1	0	15	0	1	0.9	0.9
THA-KOR	0	8	7	0	2	0.4	0.5
THA-NLD	0	1	11	1	4	0.5	0.5
THA-NOR	0	0	14	0	3	0.7	0.8
THA-USA	0	5	12	0	0	0.6	0.7
THA-VNM	0	4	3	1	3	0.3	0.2
N=223	3	53	134	6	27		
%	1.3	23.8	60.1	2.7	12.1		
USA-CAN	0	0	0	17	0	1.0	1.0
USA-CHL	0	0	3	0	14	0.7	0.8
USA-CHN	0	0	8	0	9	0.5	0.7
USA-DEU	1	3	6	2	5	0.3	0.3
USA-DNK	0	8	8	0	1	0.4	0.6
USA-ESP	3	2	10	0	2	0.6	0.6
USA-FRA	0	8	9	0	0	0.5	0.7
USA-GBR	1	1	14	0	1	0.8	0.8
USA-ISL	0	1	5	2	9	0.4	0.4
USA-ITA	0	5	10	1	1	0.4	0.5
USA-JPN	0	0	17	0	0	1.0	1.0
USA-KOR	0	1	5	1	10	0.4	0.5
USA-NLD	1	0	0	4	12	0.6	0.7
USA-NOR	2	3	12	0	0	0.7	0.7
USA-THA	0	0	12	0	5	0.6	0.7
USA-VNM	1	1	1	2	6	0.4	0.5
N=266	9	33	120	29	75		
%	3.4	12.4	45.1	10.9	28.2		

to be continued…

Table A 2: continued

Country-pair	cases					HI	MADU-0-1
	0	1	2	3	4		
VNM-CAN	1	3	0	3	0	0.5	0.7
VNM-CHL	0	3	0	1	0	0.6	0.7
VNM-CHN	0	0	0	2	9	0.7	0.8
VNM-DEU	0	1	0	1	3	0.4	0.5
VNM-DNK	0	4	3	0	0	0.5	0.7
VNM-ESP	1	2	0	0	4	0.6	0.7
VNM-FRA	0	3	4	0	1	0.4	0.5
VNM-GBR	0	1	0	2	2	0.4	0.4
VNM-ISL	0	0	0	1	3	0.6	0.7
VNM-ITA	0	0	1	1	2	0.4	0.3
VNM-JPN	0	0	2	0	9	0.7	0.8
VNM-KOR	0	2	8	0	1	0.6	0.6
VNM-NLD	0	3	2	0	2	0.3	0.3
VNM-NOR	0	6	2	3	0	0.4	0.4
VNM-THA	0	3	3	1	4	0.3	0.2
VNM-USA	1	6	1	2	1	0.4	0.5
N=124	3	37	26	17	41		
%	2.4	29.8	21.0	13.7	33.1		

67

Table A 3: Absolute frequencies of cases 0 to 4 with a tolerance limit of ± 10 percent by country pairs and their Herfindahl and MADU-0-1 indices for fresh or chilled salmonidae (SITC-code 03412), 1992-2008

Country-pair	cases 0	1	2	3	4	HI	MADU-0-1
CAN-GBR	0	3	1	4	1	0.3	0.4
CAN-NOR	0	0	0	1	0	1.0	1.0
CAN-USA	11	1	0	4	1	0.5	0.6
N=27	11	4	1	9	2		
%	40.7	14.8	3.7	33.3	7.4		
DEU-DNK	1	6	0	8	2	0.4	0.5
DEU-ESP	0	0	0	12	4	0.6	0.7
DEU-FRA	0	0	0	10	6	0.5	0.7
DEU-GBR	0	1	1	10	5	0.4	0.5
DEU-NOR	0	0	2	0	3	0.5	0.7
DEU-POL	0	0	3	2	3	0.3	0.3
DEU-SWE	0	4	0	1	0	0.7	0.7
N=84	1	11	6	43	23		
%	1.2	13.1	7.1	51.2	27.4		
DNK-DEU	1	2	0	8	6	0.4	0.5
DNK-ESP	0	5	9	1	1	0.4	0.5
DNK-FRA	0	0	0	10	7	0.5	0.7
DNK-FRO	0	0	2	0	3	0.5	0.7
DNK-GBR	0	1	2	2	11	0.5	0.6
DNK-NOR	0	5	0	12	0	0.6	0.7
DNK-POL	0	0	1	1	5	0.6	0.6
DNK-SWE	1	0	3	10	3	0.5	0.5
DNK-USA	0	4	0	1	1	0.5	0.6
N=118	2	17	17	45	37		
%	1.7	14.4	14.4	38.1	31.4		
ESP-DEU	0	4	0	12	0	0.6	0.7
ESP-DNK	0	1	9	1	5	0.4	0.5
ESP-FRA	1	9	2	3	2	0.4	0.4
ESP-GBR	0	2	2	3	4	0.3	0.2
ESP-NOR	0	1	0	0	0	1.0	1.0
ESP-SWE	0	1	0	0	0	1.0	1.0
ESP-USA	0	0	1	0	0	1.0	1.0
N=63	1	18	14	19	11		
%	1.6	28.6	22.2	30.2	17.5		

Country-pair	cases 0	1	2	3	4	HI	MADU-0-1
GBR-CAN	0	1	1	4	3	0.3	0.4
GBR-DEU	0	5	1	10	1	0.4	0.5
GBR-DNK	0	11	2	2	1	0.5	0.6
GBR-ESP	0	4	2	3	2	0.3	0.2
GBR-FRA	1	8	3	2	3	0.3	0.3
GBR-FRO	0	0	1	0	2	0.6	0.7
GBR-NOR	0	3	0	7	0	0.6	0.7
GBR-POL	0	0	1	0	0	1.0	1.0
GBR-SWE	0	5	0	1	0	0.7	0.8
GBR-USA	0	7	3	4	2	0.3	0.3
N=106	1	44	14	33	14		
%	0.9	41.5	13.2	31.1	13.2		
NOR-CAN	0	0	0	1	0	1.0	1.0
NOR-DEU	0	3	2	0	0	0.5	0.7
NOR-DNK	0	0	0	12	5	0.6	0.7
NOR-ESP	0	0	0	0	1	1.0	1.0
NOR-FRA	0	1	0	0	0	1.0	1.0
NOR-FRO	0	1	0	0	0	1.0	1.0
NOR-GBR	0	0	0	7	3	0.6	0.7
NOR-SWE	0	12	5	0	0	0.6	0.7
N=53	0	17	7	20	9		
%	0.0	32.1	13.2	37.7	17.0		
PAN-USA	0	0	0	3	0	1.0	1.0
N=3	0	0	0	3	0		
%	0.0	0.0	0.0	100.0	0.0		
POL-DEU	0	3	3	2	0	0.3	0.3
POL-DNK	0	5	1	1	0	0.6	0.6
POL-FRA	0	1	3	0	1	0.4	0.5
POL-GBR	0	0	1	0	0	1.0	1.0
POL-SWE	0	1	0	0	0	1.0	1.0
N=22	0	10	8	3	1		
%	0.0	45.5	36.4	13.6	4.5		

to be continued…

Table A 3: continued

Country-pair	cases 0	1	2	3	4	HI	MADU-0-1		Country-pair	cases 0	1	2	3	4	HI	MADU-0-1
FRA-DEU	0	6	0	10	0	0.5	0.7		SWE-DEU	0	0	0	1	4	0.7	0.7
FRA-DNK	0	7	0	10	0	0.5	0.7		SWE-DNK	1	3	3	10	0	0.5	0.5
FRA-ESP	2	1	2	3	9	0.4	0.5		SWE-ESP	0	0	0	0	1	1.0	1.0
FRA-GBR	1	3	3	2	8	0.3	0.3		SWE-FRA	0	0	0	1	5	0.7	0.8
FRA-NOR	0	0	0	0	1	1.0	1.0		SWE-GBR	0	0	0	1	5	0.7	0.8
FRA-POL	0	1	3	0	1	0.4	0.5		SWE-NOR	0	0	5	0	12	0.6	0.7
FRA-SWE	0	5	0	1	0	0.7	0.8		SWE-POL	0	0	0	0	1	1.0	1.0
FRA-USA	0	1	0	0	0	1.0	1.0		N=53	1	3	8	13	28		
N=80	3	24	8	26	19				%	1.9	5.7	15.1	24.5	52.8		
%	3.8	30.0	10.0	32.5	23.8											
									USA-CAN	11	1	0	4	1	0.5	0.6
FRO-DNK	0	3	2	0	0	0.5	0.7		USA-DNK	0	1	0	1	4	0.5	0.6
FRO-GBR	0	2	1	0	0	0.6	0.7		USA-ESP	0	0	1	0	0	1.0	1.0
FRO-NOR	0	0	0	0	1	1.0	1.0		USA-FRA	0	0	0	0	1	1.0	1.0
N=9	0	5	3	0	1				USA-GBR	0	2	3	4	7	0.3	0.3
%	0.0	55.6	33.3	0.0	11.1				USA-PAN	0	0	0	3	0	1.0	1.0
									N=44	11	4	4	12	13		
									%	25.0	9.1	9.1	27.3	29.5		

Table A 4: Average yearly im- and exports between 2005 and 2008 of fresh or chilled salmonidae (SITC-code 03412)[mio US $]

Rank	Country	Average import value [mio. US $]	Share of total world imports [%]	Country	Average export value [mio. US $]	Share of total world exports [%]
1	Sweden	916	21.9%	Norway	2,182	48.7%
2	USA	488	11.6%	Sweden	801	17.9%
3	France	473	11.3%	Canada	426	9.5%
4	Denmark	267	6.4%	United Kingdom	233	5.2%
5	Germany	251	6.0%	Denmark	169	3.8%
6	Poland	242	5.8%	Germany	130	2.9%
7	Russian Federation	216	5.2%	Chile	94	2.1%
8	Spain	177	4.2%	Panama	63	1.4%
9	Japan	171	4.1%	USA	60	1.3%
10	United Kingdom	139	3.3%	Faeroe Isds	56	1.2%
	Subtotal	3,340	79.7%	Subtotal	4,213	94.1%
	World total	4,190		World total	4,477	

Data source: UN Comtrade (2010)

Table A 5: List of country abbreviations (ISO3-digit Alpha)

Country	Abbreviation
Canada	CAN
Chile	CHL
China	CHN
Denmark	DNK
Faeroe Islands	FOR
France	FRA
Germany	DEU
Iceland	ISL
Italy	ITA
Japan	JPN
Netherlands	NLD
Norway	NOR
Panama	PAN
Poland	POL
Rep. of Korea	KOR
Russian Federation	RUS
Spain	ESP
Sweden	SWE
Thailand	THA
United Kingdom	GBR
USA	USA
Viet Nam	VNM

Source: UN (2010)

4. Fish in the network – network analysis of international fish trade

4 Fish in the network

Network analysis of international fish trade

Abstract

Fish is a highly traded product. International trade with fish increased significantly in the last decades and the pattern of fish supply changed from fisheries alone to fisheries and aquaculture. Network analysis may be a useful tool for analyzing the structure of global trade networks and to characterize the role of individual countries in a network. Trade networks of cod, salmonidae and shrimps are analyzed for the period from 1990 to 2009. The results show a clear trend towards more interlinked countries, especially for salmonidae and shrimps. More countries trade with ever more countries. The role of intermediaries in fish trade diminishes. Some network measures indicate that the exports of all three species concentrate on ever fewer but highly important countries.

1 Introduction

Fish is known as healthy food and global consumption of fish is predicted to increase (FAO 2012, FAO/WHO 2011, MOZZAFARIN and RIMM 2006). Population growth and an increasing consumption per capita are the main drivers of fish demand growth (FAO 2012). The average annual per capita fish consumption grew from 9.9 kg in the 1960s to 18.4 kg in 2009 (FAO 2012). On the supply side, capture fisheries were the only source of fish for a long time. But the pattern has changed as total fishery production stagnates since the end of the 1980s and aquaculture production became the fastest growing food sector of the world and accounts today for nearly 50 percent of total food fish supply (FAO 2012).

Growing demand for fish with simultaneous stagnating fisheries production, trade liberalization, outsourcing of processing and improvements in transporting and cooling technologies have led to an increasing international trade with fish. The share of production entering international trade increased from 25 percent in 1976 to 38 percent in 2010 (FAO 2012). Trade with fish accounted for 10 percent of total agricultural exports and 1 percent of world merchandise trade in value terms (FAO 2012).

Trade can be described as a network with countries as nodes and the trade flows as links connecting the nodes. A network is thus influenced by the characteristics of production and demand and the countries are connected by import and export trade flows (DE BENEDICTIS and TAJOLI 2009). The analysis of trade with tools of network analysis may offer some insights in the global structure and also on the role of each individual country. Network analysis may serve as a suitable toolbox to describe changing patterns of international trade in a quantitative way (KIM and SHIN 2002). Network-based approaches may provide a more powerful way to manage, monitor, and govern complex systems (SCHWEITZER et al. 2009).

This paper deals with a commodity based network analysis of fish trade. Three species are selected: cod, salmonidae and shrimps. Cod has a long tradition in fisheries and human consumption and is supplied by capture fisheries. The main fishery nations of cod are USA, Canada, and Northern and Western European countries. Salmonidae includes salmon and trout. Salmonidae are mainly produced in aquaculture although there are still some salmonidae caught in the wild. The main producer countries of salmon are Norway, Chile, the United Kingdom and Canada. Today, shrimps and prawns origin from capture fisheries and aquaculture with nearly equal shares. Many shrimp farms are operated in Asian countries.

The goal of this paper is to analyze the development of the trading networks of cod, salmonidae and shrimps and to examine if there are differences in trade networks between fish originating from capture fisheries, or from aquaculture, or from both. Characteristics of the markets will be identified to test if there are changes in the density and centralization of networks.

The paper is structured as follows. The next section briefly describes the data set and section 3 gives an overview on the development of production and trade of the selected species. Selected measures of network characteristics are described in Section 4. Also the results of the network analysis for valued and binarized trade networks are presented here. The paper closes with a summary and a discussion.

2 Data

The data analyzed in this study originate from UN Comtrade data base. UN Comtrade is provided by the UN and the trade data base contains more than 1.7 bn. trade records in several commodity classifications since 1962 (UN COMTRADE 2011). The Standard International Trade Classification (SITC) in its third revision is selected to analyze the international trade networks of cod, shrimps, and salmonidae between 1988 and 2010. For cod and salmonidae several four digit commodity codes exist, which were aggregated for the purposes of this study. Table 1 shows the selected SITC codes.

Table 1: Aggregation of SITC-Codes to one species for network analysis

Species	SITC-Code	Description
Cod	03416	Cod, fresh or chilled (excluding livers and roes)
	03425	Cod, frozen (excluding livers and roes)
	03511	Cod, dried, whether or not salted
	03521	Cod, salted but not dried or smoked and fish in brine
Shrimps	03611	Shrimps and prawns, frozen
Salmonidae	03412	Salmonidae, fresh or chilled (excluding livers and roes)
	03421	Salmonidae, frozen (excluding livers and roes)
	03711	Salmon, whole or in pieces, but not minced

For the network analysis only import trade data in current U.S. dollars were used, as import data are believed to be more consistent in the sense of reliability and completeness (DE BENEDICTIS and TAJOLI 2009). One often named reason is that custom offices seem to be more interested in goods that enter a country to collect tariffs (DE BENEDICTIS and TAJOLI 2009). Moreover, customs officers cannot always know where an export will eventually go, but they may be able to trace its origin. A previous study showed that fish trade data are inaccurate and that there are discrepancies between recorded import and export values of the same transaction (GUETTLER 2013, Chapter 3 in this dissertation). However, for each species the data were than transformed into $n \times n$ matrices containing the fish trade data of n import countries with n exporting partner countries.

3   Development of production and trade

This section gives a short overview of the development of the production and of the international trade of the selected fish species cod, salmonidae, and shrimps. In 1968 cod production peaked at more than 4 mio t and then fell to less than 1.2 mio t in 2009 (Figure 1). The share of cod aquaculture production is insignificant (Table 2). The production of salmon and trout (salmonidae) by capture fisheries increased from 0.4 mio t in 1950 to 1.2 mio t in 2009 while the aquaculture production reached 0.4 mio t in 1988 and increased to 2.5 mio t in 2009 (Figure 2). The share of aquaculture production of salmonidae on total salmon production grew from 39 percent in 1990 to 67 percent in 2009 (Table 2). Capture fisheries production of shrimps and prawns increased from 0.4 mio t in 1950 to 3.3 mio t in 2003 and stagnates since then while the aquaculture production still rises and reached 3.5 mio t in 2009 (Figure 3). In 2007 shrimps aquaculture production was higher than production from capture fisheries for the first time and reached a share of 52.7 percent in 2009 (Table 2).

Figure 1: Development of cod production, 1950 - 2009]

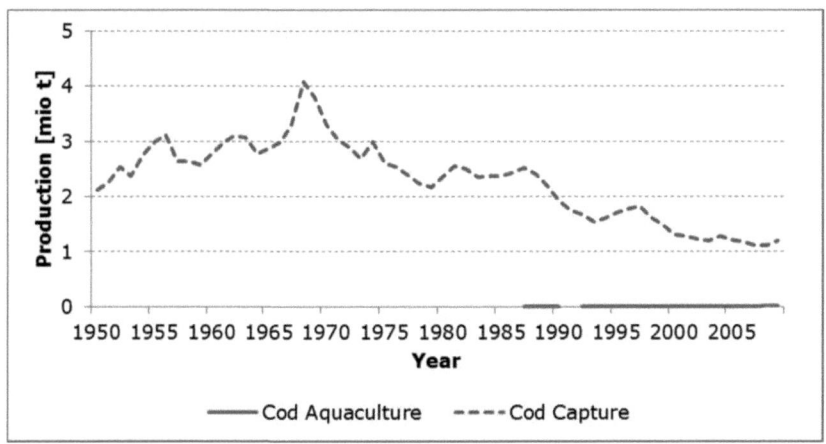

Data Source: FAO (2011)

Figure 2: Development of salmonidae production, 1950 - 2009

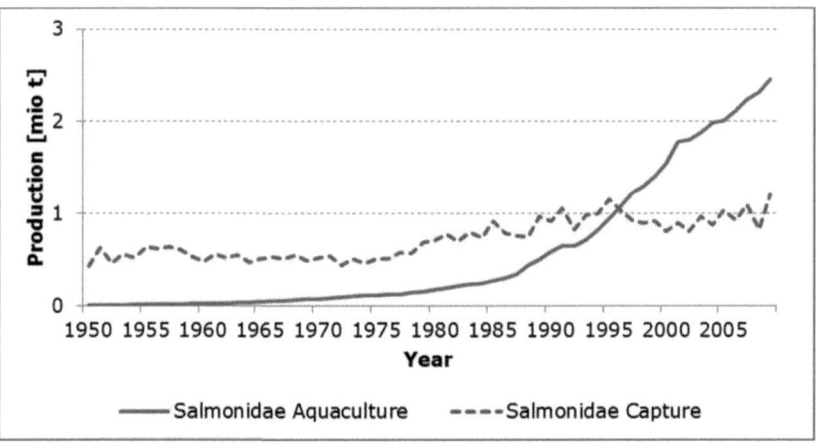

Data Source: FAO (2011)

Figure 3: Development of shrimp production, 1950 – 2009

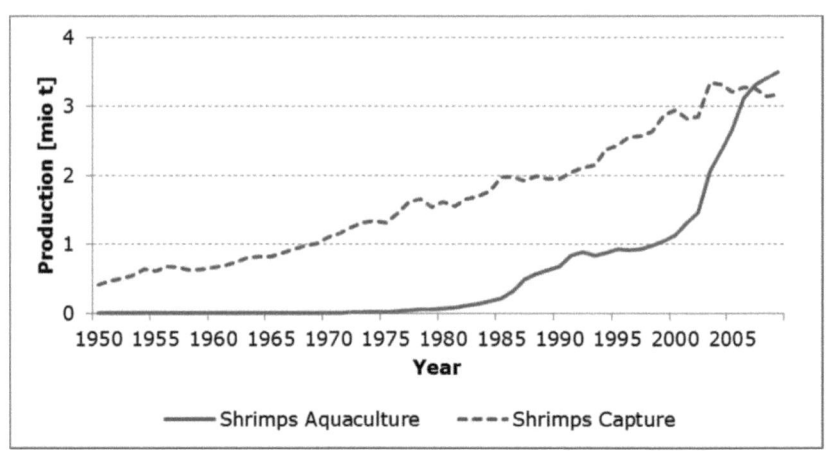

Data Source: FAO (2011)

Table 2: Percentage share of aquaculture production on total production

Species	Year				
	1990	1995	2000	2005	2009
Cod	0.0	0.0	0.0	0.7	1.9
Salmonidae	39.1	44.9	65.7	66.0	67.1
Shrimps	25.8	27.5	27.9	45.4	52.4

Data Source: FAO (2011)

Total import value of all three species increased between 1988 and 2010. Figure 4 shows the development of the total import value of the three species.

The import value of cod increased from 1 bn US$ in 1988 to 2.6 bn US$ in 2010 which equals a compound annual growth rate (CAGR) of 4.2 percent. Cod imports peaked in 2007 with an import value of nearly 3.7 bn US$.

The imports of salmon increased slowly between 1988 and 2002 as trade value rise from 2.3 bn US$ to 3.3 bn US$, but then suddenly rise to about 9.6 bn US $ in 2010. The CAGR for salmon imports between 1988 and 2010 accounts for 6.7 percent.

Import value of shrimps stood at 4 bn US$ in 1988 and increased to 9 bn US$ in 1995 and 11 bn US$ in 2011. The CAGR is 4.7 percent between 1988 and 2010.

Figure 4: Development of the value of imports of cod, salmonidae and shrimps, 1988-2010

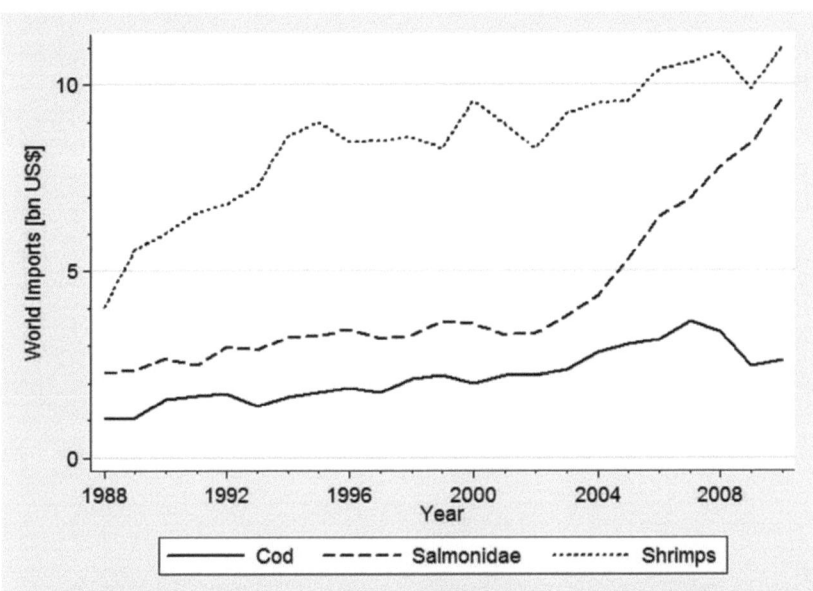

Simultaneously to the growing import values also the import quantities of the selected species have grown, which is shown in Figure 5. Trade quantities are only given for the years from 1988 to 2009, because the import quantity of cod increased to an incredible value of more than 3 mio t and the maximum quantity in the named period was "only" 1.1 mio t in 2005.

Cod trade increased by 1.7 percent per year from 0.4 mio t in 1988 to 0.6 mio t in 2009, but was around 1 mio t for a long time (1994 – 2006).

Global import quantity of salmonidae grew from 0.3 mio t to 1.8 mio t between 1988 and 2009. Compared to cod and shrimp, salmonidae had the lowest import quantity in 1988 and became the main important trade species in 2009. In 2006, salmonidae trade had reached its maximum in the analyzed period with a trade volume of 2.4 mio t. The CAGR for salmonidae import quantity between 1988 and 2009 is 8.3 percent.

The import quantity of shrimps increased continuously from 0.5 mio t in 1988 to 1.5 mio t in 2009, which equals a CAGR of 5.5 percent. For the period from

1988 to 2005 shrimps were the main important trade species by volume and value compared to cod and salmon.

Figure 5: Development of the quantity of imports of cod, salmonidae and shrimps, 1988-2009

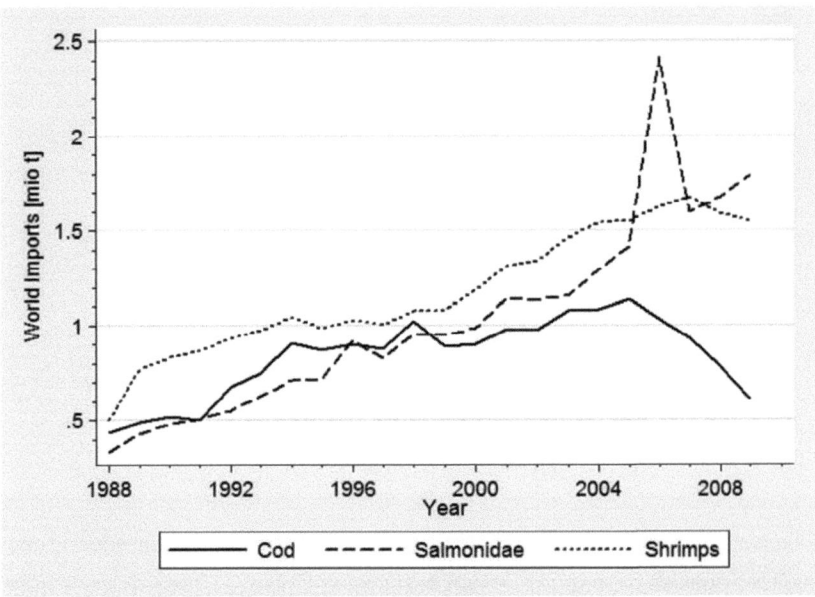

Figure 4 and Figure 5 show that trade quantities and values for cod, salmon and shrimps increased over time. These data were used to compute global import prices of these three species, which are depicted in Figure 6. While the prices for shrimps and salmonidae drop between 1988 and 2009 by 1.1 percent and 1.8 percent per year, respectively, prices of cod increased by 2.4 percent per year. Nevertheless, cod has still the lowest price of all three species. The price of 1 kg cod was 4 US$ in 2009 and 2.41 US$ in 1988.

Improvements in salmon farming technology led to decreasing production costs which are also reflected in the trade prices (ASCHE et al. 1999; ASCHE 1997). The price of salmon was 6.81 US$/kg in 1988 and sunk to 2,67 US$/kg in 2006 and then increased to 4.70 US$/kg in 2009.

Shrimps gained a much higher price on the global import market. Prices stood at 8 US$/kg in 1988, reached a maximum of 9.13 US$/kg in 1995 and then fell to 6.37 US/kg in 2009.

Figure 6: Development of import prices of cod, salmonidae and shrimps, 1988-2009

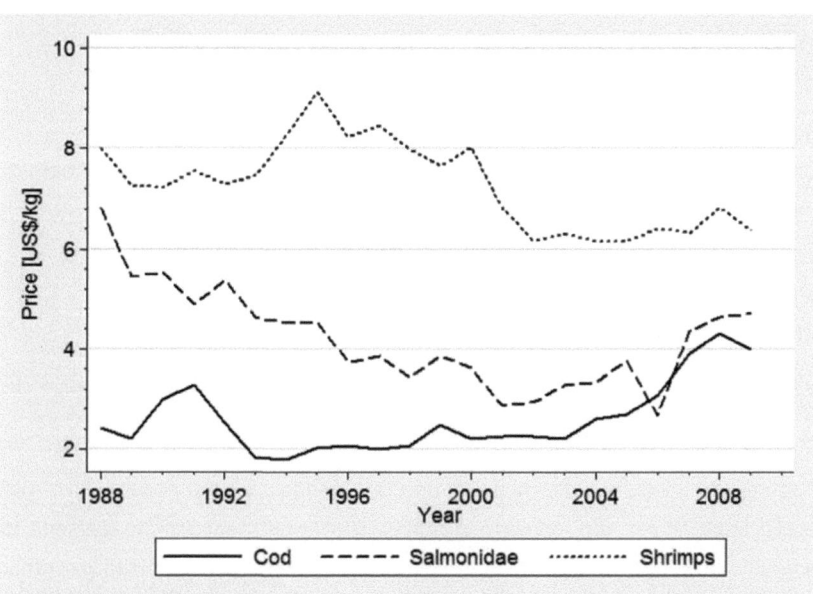

4 Networks of fish trade

4.1 Network topology

Networks are based on graph theory and some concepts of the analysis of networks rely on graph theory. A network $\mathcal{G}(\mathcal{N}, \mathcal{L})$ is defined by a set of nodes $\mathcal{N} = \{n_1, n_2, \dots, n_g\}$ and a set of edges $\mathcal{L} = \{l_1, l_2, \dots, l_L\}$. Nodes represent actors, in our case countries, and edges represent ties or relationships between the actors, in our case trade flows between trading partners. To be precise, in the case of directed graphs, also named digraph, the edges are called arcs and the ties are oriented from one actor to another. Trade networks are an excellent example for digraphs. Goods are transferred from one country to another country, so that one country is the source (the exporter) of the good, while the other nation is the destination of the good (the importer) (WASSERMAN and FAUST 1994).

Some network indices characterize the role of a single country in the trade network, e.g. the number of trading partners. This allows comparing the countries with each other. Moreover, some other network indices describe the whole network, e.g. centralization measures. Different networks can be compared by these global network indices. Selected measures of network characteristics are described briefly in the following subsections.

4.1.1 Degree

The degree d of a vertex or node n_j is the number of its connected neighbors or the number of edges (FREEMAN 1979, WASSERMAN and FAUST 1994).

$$C_D(n_j) = d(n_j) = \sum_{i=1}^{g} x_{ij} \tag{1}$$

where:

x_{ij}: = 1 if and only if n_i and n_j are connected by a line (otherwise 0), and

g: total number of nodes in the network.

The degree is an absolute measure and is dependent on the network size. For some applications, e.g. comparing nodes in different networks, it is useful to compute a relative degree. As a vertex can at most be connected to g-1 other vertices, the degree centrality C_D' of node n_i is measured as follows (FREEMAN 1979, WASSERMAN and FAUST 1994):

$$C_D'(n_i) = \frac{d(n_i)}{g-1} \tag{2}$$

While the *degree centrality* C_D' is a measurement for one actor, the *degree centralization* C_D is a measurement for the whole network and reflects the range or variability of the individual degree indices (FREEMAN 1979, WASSERMAN and FAUST 1994).

$$C_D = \frac{\sum_{i=1}^{g}[C_D(n^*) - C_D(n_i)]}{(g-1)(g-2)} \tag{3}$$

where:

$C_D(n^*)$: largest value of $C_D(n_i)$ for any node in the network.

In directed networks the direction of the arc is taken into account. It is then possible to separate the degree into an *indegree* d_I, that is the number of arcs terminating in n_i, and into an *outdegree* d_O of a node, that is the number of arcs originating with n_i. A country with a relatively high indegree is a heavy importer and a country with a high outdegree is a heavy exporter (WASSERMAN and FAUST 1994). The *outdegree centrality* C'_{DO} and *indegree centrality* C'_{DI} of a node n_i are measured as follows:

$$C'_{DO}(n_i) = \frac{d_O(n_i)}{g-1} \tag{4}$$

$$C'_{DI}(n_i) = \frac{d_I(n_i)}{g-1} \tag{5}$$

Outdegree and *indegree centralization* are measured similar to degree centralization:

$$C_{DO} = \frac{\sum_{i=1}^{g}[C_{DO}(n^*) - C_{DO}(n_i)]}{(g-1)(g-2)} \tag{6}$$

$$C_{DI} = \frac{\sum_{i=1}^{g}[C_{DI}(n^*) - C_{DI}(n_i)]}{(g-1)(g-2)} \tag{7}$$

While the out- or indegree are measurements for each single node, the *mean outdegree* $\overline{d_O}$ and *mean indegree* $\overline{d_I}$ are global values, describing the whole network, which can be measured as (FREEMAN 1979, WASSERMAN and FAUST 1994):

$$\overline{d_I} = \frac{\sum_{i=1}^{g} d_I(n_i)}{g} \tag{8}$$

$$\overline{d_O} = \frac{\sum_{i=1}^{g} d_O(n_i)}{g} \tag{9}$$

It can be shown that $\overline{d_O} = \overline{d_I}$, because each arc is counted twice: as an ingoing and an outgoing arc, and thus the sum of all indegrees equals the sum of all outdegrees. But the variance S^2 of the in- and outdegree is not necessarily the same:

$$S^2_{d_I} = \frac{\sum_{i=1}^{g}(d_I(n_i) - \overline{d_I})^2}{g} \tag{10}$$

$$S^2_{d_O} = \frac{\sum_{i=1}^{g}(d_O(n_i) - \overline{d_O})^2}{g} \tag{11}$$

The variance of the in- and outdegree quantifies, how unequal the actors in a network are with respect to initiating or receiving ties (WASSERMAN and FAUST 1994).

In the analysis of fish trade networks the calculation of the indegree (outdegree) allows to identify the main importers (exporters) by their number of direct trading partners and trading values and thus have an influential part on the trade of fish. Indegree and outdegree centralization help to compare different networks over time and fish species and show how centralized these networks are.

4.1.2 Density

The density Δ is a measurement based on the degree and shows how many edges or arcs are existent in a network compared to the maximum possible number of edges or arcs. Each node could be connected to g-1 nodes, as long as we assume that there are no loops. This limits the maximum to $g(g$-1$)$ possible edges or arcs in a network. The minimum density is 0 if no arcs are present and the maximum density is 1 if all arcs are present and each node is connected with each other node (WASSERMAN and FAUST 1994). The density for a directed network is computed by

$$\Delta = \frac{\sum_{i=1}^{g} d_I(n_i)}{g(g-1)} = \frac{\sum_{i=1}^{g} d_O(n_i)}{g(g-1)} \tag{12}$$

In relation to fish trade networks, the density shows how dense a network is or how many trading partnerships are existent compared to the maximum possible number of partnerships. This gives some information on the structure of different networks.

4.1.3 Betweenness Centrality

The idea behind the network measurement *betweenness centrality* is that an actor is central if it lies between two actors on their *geodesic*. A *geodesic* is the shortest path between two actors or nodes and the *geodesic distance d(i,j)* is the length of the *geodesic* between these pair of nodes n_i and n_j (WASSERMAN and FAUST 1994). The actor on the shortest path might have some control over the interaction between the two actors, which are not connected directly. In contrast to the degree, the betweenness centrality quantifies how an actor is connected to the rest of the network rather than just counting the direct connections. The *actor betweenness centrality $C_B(n_i)$* for n_i is computed as (FREEMAN 1979, GOULD 1987, WASSERMAN and FAUST 1994):

$$C_B(n_i) = \sum_{j<k} g_{jk}(n_i)/g_{jk} \tag{13}$$

where:

g_{jk}:   number of geodesics linking the actors j and k, and

$g_{jk}(n_i)$:   number of geodesics connecting the two actors that contain actor i.

In a digraph or non-symmetric networks an actor can lie on a maximum of $(g-1)(g-2)$ geodesics (GOULD 1987). Therefore the betweenness centrality for non-symmetric networks can be standardized to values between 0 and 1 by:

$$C_B'(n_i) = \frac{C_B(n_i)}{(g-1)(g-2)}$$

(14)

The *standardized betweenness centrality index* C_B' can easily be compared to other actor based indices or across networks or relations. One advantage of the betweenness centrality is that it can be computed, even if the graph is not connected (WASSERMAN and FAUST 1994). This is not the case for the *closeness centrality*, which is based on the distance of one actor to all other actors and cannot be computed reasonably if isolated actors exist and thus no distance can be measured.

To compare different networks with respect to the heterogeneity of the betweenness of the actors of a whole network, it is useful to measure the *betweenness centralization* C_B. C_B takes the value 0 for all networks of any size if the centrality of all nodes is equal and C_B is 1 in the case of a star-shaped network (FREEMAN 1977, 1979, WASSERMAN and FAUST 1994).

$$C_B = \frac{2\sum_{i=1}^{g}[C_B(n^*)-C_B(n_i)]}{[(g-1)^2(g-2)]} = \frac{\sum_{i=1}^{g}[C_B'(n^*)-C_B'(n_i)]}{(g-1)}$$

(15)

where:

$C_B(n^*)$:   the largest realized actor betweenness index for the set of actors,

$C_B'(n^*)$: the largest realized standardized actor betweenness index for the set of actors.

GOULD (1987) argues that the computation of betweenness centralization might be problematic in the case of non-symmetric networks, because the removal of one arc does not necessarily lead to a decline in the value of C_B, which is the case for symmetric networks.

In the analysis of international fish trade the betweenness centrality may be an indicator of a countries influence of the network. A relatively high betweenness centrality could be useful to identify intermediaries or distributors in the fish supply chain, which are countries which import fish, process it eventually and export it afterwards. So, these intermediaries may have an influence on the trade flow of fish in the world.

4.2 Literature overview

Many publications are dealing with trade and networks, but only a few really apply network analysis to real trade data. Recent publications are DE BENEDICTIS and TAJOLI (2009, 2011), and CASSI et al. (2009) which analyze trade networks and compute some specific network parameters for different commodity groups. However, it seems that fish trade networks were not analyzed yet.

DE BENEDICTIS and TAJOLI (2009) analyzed trade networks of 28 commodity groups for the year 2000. For example, the networks of food products, beverages, tobacco were examined but also the networks of petroleum, iron and steel, and machinery were scrutinized. Import data were used to compute most of the network parameters presented in the previous section. Cores in the network, which is a relatively dense sub-network within the network, are identified for each of the products, too. DE BENEDICTIS and TAJOLI (2009) conclude that the characteristics of the trade networks display remarkable differences and that these differences are based on the characteristics of a good.

The world trade network is analyzed by DE BENEDICTIS and TAJOLI (2011). Aggregated import data are used to compute the network indices of the previous section for the period from 1950 to 2000. The main important countries based on different centrality measures are presented. DE BENEDICTIS and TAJOLI (2011) find that the trading system has become more interconnected, while the heterogeneity between the countries increased over time. Moreover, the authors conclude that trade policies had an impact on the structure of the networks.

CASSI et al. (2009) analyzed wine trade networks for a limited number of 24 countries and used cut-off values of 1 and 2 mio US $. Indegree, outdegree and the density of the wine trade network are computed for the period from 1974 to 2004. Additionally, block modeling techniques were used to distinct countries in core exporters and core consumers. CASSI *et al.* (2009) conclude that the participation of countries as importers and exporters increased over time.

4.3 Results

In this analysis fish trade networks are analyzed from two perspectives: i) trading value and ii) number of trading partnerships. UN Comtrade data were used to create the matrices, which were analyzed with UCINET 6 (BORGATTI *et al.* 2002).

4.3.1 Networks by trading value

The basic characteristics of the trade networks of cod, salmonidae and shrimps are presented in Table 3. Total trade value increased for all three species between 1990 and 2009 (see Figure 4 for details). The number of countries participating in the trade of cod, salmon and shrimps also increased over time, which might also be a result of an increase in the number of countries in the world. While about 170 to 180 countries import or export salmon and shrimps, the number of countries trading cod is lower and reached a maximum of 154 countries in 2005. With increasing trade values and an increasing number of participating countries it is not astonishing that more trade connections (number of arcs) emerged. The number of trading connections increased between 1990 and 2009 by 70 percent for cod, 115 percent for salmon, and 55 percent for shrimps. Due to the increasing number of countries, the density of the networks was more or less constant or decreased in some years. The development of the average trade value between the countries is different for the three species. For cod and salmon the average value decreased between 1990 and 2000 and increased afterwards. But the average trade value in 2009 is lower for cod and higher for salmon compared to the value in 1990. The average trade value for shrimps fluctuates between 5.1 and 6.6 mio. US\$ and the value in 2009 nearly equals the value in 1990. The standard deviation is lowest for cod while the trade with salmon and shrimps shows significant higher standard deviations. This indicates that the difference between large and small importers or exporters is more distinct in the salmon and shrimp trade networks.

Table 3: Basic characteristics of the trade networks

Species	Year	Total trade value [bn. US $]	Number of countries (nodes)	Number of arcs	Density [%]	Avg. trade value (over arcs) [mio. US $]	Std. Dev. [mio. US $]
Cod	1990	1.554	84	392	5.6	4.0	11.9
	1995	1.270	122	524	3.5	2.4	11.0
	2000	1.314	149	622	2.8	2.1	8.8
	2005	1.854	154	726	3.1	2.6	11.3
	2009	2.018	128	665	4.1	3.0	13.5
Salmon	1990	2.657	109	669	5.7	4.0	28.9
	1995	3.170	147	999	4.7	3.2	22.9
	2000	3.395	178	1260	4.0	2.7	21.1
	2005	4.971	178	1472	4.7	3.4	26.3
	2009	7.892	168	1437	5.1	5.5	46.0
Shrimps	1990	6.023	145	1118	5.4	5.4	30.6
	1995	8.890	156	1349	5.6	6.6	45.1
	2000	9.255	174	1575	5.2	5.9	38.3
	2005	9.284	179	1812	5.7	5.1	29.0
	2009	9.580	170	1733	6.0	5.5	32.0

Figure 7 shows the trade networks of salmonidae of the years 1990 and 2009. The number of countries and the number of trade connections increased. By looking at Figure 7 one might guess that the density of the network was significantly higher in 2009 than in 1990. A look at Table 3 reveals that the opposite is the case: The density in 1990 was 5.7 percent compared to 5.1 percent in 2009. The increasing number of countries led to an over proportional increase of the maximum possible number of links. As the number of trade connections did not rise at that rate, the ratio of existent arcs to the maximum possible number of arcs (density) declined. Graphical representations of networks cannot easily be visually inspected and may lead to misleading conclusions (SCOTT 2000). Although there are some general principles for network visualization, the representation of network figures may occur randomly (SCOTT 2000). Thus, the explanatory power of such graphs is limited. Therefore, in the following characteristics of the networks will only be presented by network indices.

Figure 7: Graphical representation of the salmonidae trade network, 1990 and 2009

1990   2009

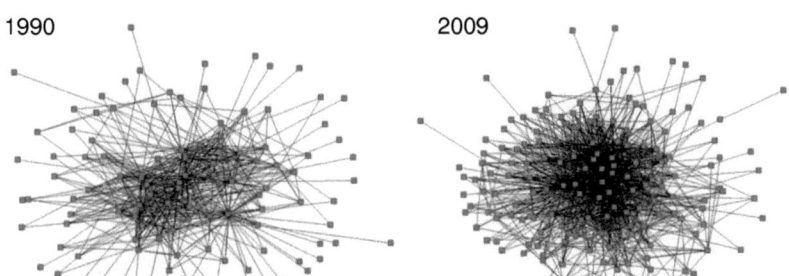

Some properties of the whole trade networks are presented in Table 4. The mean indegree or mean outdegree, which is the mean trade value of each country, increased for shrimps and salmon between 1990 and 2009, while it decreased for cod. The mean trade value per country is highest for shrimps, but a strong growth could be detected for salmon. The mean import or export values nearly doubled between 1990 and 2009. The average trade value for cod dropped by 10 mio. US$ between 1990 and 2000 but then recovered and reached nearly 16 mio US$ in 2009. In 1990 the maximum imports are higher than the maximum exports for all three species. In the following years the maximum exports are larger than the imports for cod and salmon, while for shrimps the maximum imports are larger than the maximum exports. The standard deviation of the trade value for salmon is higher for exports than for imports. This indicates that the importing countries are more similar in terms of the import value than it is the case for exporting countries. The same holds for the trade with cod since 1995. The opposite is the case for trade with shrimps: The exporting countries are more similar in terms of the export value than it is the case for the importing countries.

All analyzed networks show low degree centralization. This indicates that all countries are in a similar position and have similar number of trading partners. The outdegree centralization for cod and salmon is a little bit higher than the indegree centralization for the years from 1995 to 2009. This indicates that the differences between the trade values of the exporting countries are a little bit larger than for the importing countries. The opposite is the case for shrimps.

Table 4: Network indices of trade networks by value

Species	Year	Mean Indegree (Out-degree) $\bar{d}_I = \bar{d}_O$ [mio. US $]	Max. [mio. US $]		Std. dev. [mio. US $]		Inde-gree Central-ization	Outde-gree Central-ization
			Inde-gree d_I	Outde-gree d_O	Inde-gree d_I	Outde-gree d_O	C_{DI} [%]	C_{DO} [%]
Cod	1990	18.5	388.5	284.8	55.6	54.4	4.6	3.3
	1995	10.4	351.1	441.2	38.0	48.1	1.5	1.9
	2000	8.8	304.8	360.8	31.8	39.5	1.8	2.1
	2005	12.0	403.8	470.2	45.8	52.1	1.3	1.5
	2009	15.8	361.9	623.0	48.2	65.6	1.1	2.0
Salmon	1990	24.4	969.8	863.2	106.5	116.5	1.4	1.2
	1995	21.6	1006.1	1047.7	98.7	114.5	1.6	1.7
	2000	19.1	990.9	1254.5	89.4	112.8	1.2	1.5
	2005	27.9	920.4	2232.2	103.7	181.9	0.8	2.1
	2009	46.9	1336.8	3835.2	162.1	313.4	0.6	1.7
Shrimps	1990	41.5	2486.2	775.7	247.6	118.4	3.4	1.0
	1995	57.0	3459.3	1979.5	340.7	201.5	2.5	1.4
	2000	53.2	3122.1	1547.8	317.1	175.2	1.8	0.9
	2005	51.8	2920.2	1095.7	272.1	161.6	2.6	1.0
	2009	56.3	2896.2	1455.5	270.4	189.3	2.1	1.0

Table 5, Table 6 and Table 7 show the top five importing and exporting countries of the selected years by trade value. For all three species the main exporting and importing countries do not change much over the years. For example, in the case of cod Norway, Iceland and the USA belong to the top five exporting countries over all five points of time and for the importing countries this is true for Portugal and Spain. Norway increased its exports of cod significantly between 1990 and 2009, while the exports of the other countries remained on a lower level.

Table 5: Degree centrality of cod trade networks by value

Year	Rank	Country	Outdegree Centralization C_{DO} (n_i) [mio. US $]	Country	Indegree Centralization C_{DI} (n_i) [mio. US $]
1990	1	Norway	285	Portugal	389
(n=84)	2	Iceland	264	Denmark	230
	3	Denmark	214	Italy	152
	4	USA	191	Spain	126
	5	Canada	164	United Kingdom	122
1995	1	Norway	441	Portugal	351
(n=122)	2	Denmark	240	Spain	130
	3	Iceland	138	Italy	99
	4	USA	117	Brazil	97
	5	Canada	45	France	93
2000	1	Norway	361	Portugal	305
(n=149)	2	Iceland	218	Spain	123
	3	USA	168	France	104
	4	Denmark	151	Italy	94
	5	Netherlands	63	China	78
2005	1	Norway	470	Portugal	404
(n=154)	2	USA	271	Sweden	211
	3	Denmark	243	Spain	206
	4	Iceland	159	China	192
	5	Sweden	155	Italy	132
2009	1	Norway	623	Portugal	362
(n=128)	2	USA	216	Sweden	250
	3	Iceland	197	Spain	172
	4	Netherlands	190	China	162
	5	Sweden	178	France	125

In the case of shrimps (Table 6), the main exporting countries are Asian countries, in particular Thailand, Vietnam, and Indonesia. The export values show also the emergence of Vietnam as an exporter of shrimps. The main importing countries of shrimps are Japan, USA, and European countries, in particular Spain, France, and Italy.

Table 6: Degree centrality of shrimp trade networks by value

Year	Rank	Country	Outdegree Central- ization C_{DO} (n_i) [mio. US $]	Country	Indegree Central- ization C_{DI} (n_i) [mio. US $]
1990	1	China	776	Japan	2486
(n=145)	2	Thailand	757	USA	1589
	3	Indonesia	648	Spain	449
	4	Ecuador	417	France	287
	5	India	391	Denmark	214
1995	1	Thailand	1980	Japan	3459
(n=156)	2	Indonesia	1008	USA	2416
	3	Ecuador	785	Spain	606
	4	India	779	France	410
	5	Mexico	374	Denmark	224
2000	1	Thailand	1548	USA	3122
(n=174)	2	Indonesia	1008	Japan	2677
	3	India	1003	Spain	748
	4	Viet Nam	617	France	374
	5	China	458	United Kingdom	261
2005	1	Viet Nam	1096	USA	2920
(n=179)	2	Indonesia	968	Japan	1837
	3	Thailand	965	Spain	1015
	4	India	891	France	533
	5	China	582	Italy	342
2009	1	Thailand	1455	USA	2896
(n=170)	2	Viet Nam	1120	Japan	1643
	3	Indonesia	917	Spain	938
	4	India	799	France	561
	5	Ecuador	786	Italy	390

The export of salmon is dominated by the largest aquaculture producer of these species, namely Norway. Table 7 shows also the emergence of Chile, which became the second largest exporter of salmon. Canada and the USA belong to the top five exporting countries at all points of time, but both countries became less important over time in relation to Norway and Chile. The main importing countries in terms of trade value are Japan, USA, Sweden, France and, in 2009, Russia.

Table 7: Degree centrality of salmonidae trade networks by value

Year	Rank	Country	Outdegree Central-ization C_{DO} (n_i) [mio. US $]	Country	Indegree Central-ization C_{DI} (n_i) [mio. US $]
1990	1	USA	863	Japan	970
(n=109)	2	Norway	761	France	409
	3	Canada	412	USA	301
	4	Chile	139	United Kingdom	185
	5	Denmark	123	FFR Germany	148
1995	1	Norway	1048	Japan	1006
(n=147)	2	USA	732	France	400
	3	Chile	413	Germany	319
	4	Canada	357	USA	307
	5	Denmark	204	Denmark	210
2000	1	Norway	1255	Japan	991
(n=178)	2	Chile	554	USA	373
	3	USA	466	France	338
	4	Canada	380	Germany	280
	5	Denmark	201	Sweden	226
2005	1	Norway	2232	Japan	920
(n=178)	2	Chile	645	Sweden	621
	3	USA	520	France	472
	4	Canada	464	USA	438
	5	Denmark	259	Germany	324
2009	1	Norway	3835	Sweden	1337
(n=168)	2	Chile	1069	Japan	1021
	3	USA	659	USA	697
	4	Canada	480	France	622
	5	Sweden	309	Russian Federation	570

4.3.2 Binary trade networks

In the previous section the international trade of cod, salmon and shrimps was analyzed by trade value. This section uses the number of trade connections to gain additional information on the trade networks of these three species. Therefore the initial trade value matrices were binarized, using only the values one and zero. If $x_{ij}=1$ in the trade matrix, there is a trade connection between country i and j, and if $x_{ij}=0$ there is no trade between both countries. The network indices of the binary trade networks for cod, salmon and shrimps are presented in Table 8.

The density of the networks, which is the number of arcs present compared to the maximum possible number of arcs, lies between 3 percent and 6 percent and is therefore relatively low but more or less constant over time. The growing number of participating countries led to an over proportional increase of possible connections. The more or less constant density indicates that the increasing number of trade connections (arcs) equalized the increasing number of possible arcs (compare Table 3).

On average, each country trades cod with 4 to 5 countries, salmon with 6 to 9 countries and shrimps with 7 to 10 countries. For this three species, the average number of trading partners increased over time. The maximum number of trade connection for imports varies between cod, salmon and shrimps. The number is lowest for cod (23 to 28) and highest for shrimps (63 to 89). Also the maximum number of exporting connections increased over time: for cod from 29 in 1990 to 61 in 2009, for salmon from 34 in 1990 to 102 in 2009 and for shrimps from 30 in 1990 to 77 in 2009.

The development of the median indegree and median outdegree is contrary over time. While the median indegree increases significantly over time for all three species, the median outdegree stays constant for cod and drops for salmon and shrimps. This indicates that countries tend to import fish from an increasing number of trading partners. The decreasing import standard deviation between 1990 and 2009 indicates the importing countries became more similar in their numbers of trading partners. The decreasing median outdegree indicates that there seems to be many countries which have none or only one export trading partner. The outdegree standard deviation (not shown in Table 8) increased between 1990 and 2009. Thus exporting countries became more different in the number of trading partners.

The degree centralization indices give information on the differences of degrees between the countries in a network. As the indegree centralization decreased over time, it can be concluded that the difference between the importing country with the highest indegree centrality and the other countries decreased. The importing countries became more similar in the number of trading partners. The outdegree centralization increased over time and the average gap between large and small exporting countries measured by the number of trading partners rose. The development of the centralization indices is in line with the development of the median indegree and median outdegree and the development of their standard deviation. For cod, the outdegree centralization is larger than the indegree for all points of time, while this is the case for salmon since 1995 and for shrimps since 2005.

The betweenness centralization is relatively low for all species. This indicates that the relevance of intermediaries or hubs is low. Fish seems to be mainly imported from the country where the fish was caught or produced.

Table 8: Network indices of binary trade networks

			Mean Inde-gree (Outde-gree)	Max.		Median		Inde-gree Central-ization	Outde-gree Central-ization	Between-ness Cen-trality
Species	Year	Density		Inde-gree	Outde-gree	Inde-gree	Outde-gree			
		Δ [%]	$\bar{d}_I = \bar{d}_O$	d_I	d_O	d_I	d_O	C_{DI} [%]	C_{DO} [%]	C_B [%]
Cod	1990	5.6	4.7	27	29	1.5	1	27.2	29.7	8.0
	1995	3.6	4.3	23	51	3	1	15.6	38.9	8.4
	2000	2.8	4.2	23	55	2	1	12.8	34.6	6.1
	2005	3.1	4.7	28	62	2	1	15.3	37.7	7.4
	2009	4.1	5.2	23	61	3	1	14.1	44.3	6.9
Salmon	1990	5.7	6.1	42	34	0	2	33.5	26.0	9.2
	1995	4.7	6.8	33	84	4	2	18.1	53.2	10.5
	2000	4.0	7.1	38	90	4.5	1	17.6	47.1	12.1
	2005	4.7	8.3	34	108	6	1	14.6	56.7	8.8
	2009	5.1	8.6	36	102	7	1	16.5	56.3	7.1
Shrimps	1990	5.4	7.7	89	30	0	5	56.8	15.6	5.1
	1995	5.6	8.6	73	51	2	4	41.8	27.5	8.1
	2000	5.2	9.1	76	58	4	2	38.9	28.5	11.6
	2005	5.7	10.1	66	81	5	2	31.6	40.0	8.1
	2009	6.0	10.2	63	77	7	2	31.4	39.8	7.4

Table 9, Table 10, and Table 11 show the five largest importing and exporting countries by the number of trading partners for cod, shrimps, and salmon respectively. Table 9 shows that Norway exports cod to most other countries. For example, in 2009 Norway exported cod to nearly each second country of the remaining 127 countries. The indegree centrality of the top five importing countries decreases between 1990 and 2000 and slightly increases afterwards as it has also been detected for the indegree centralization.

Table 9: Degree centrality of cod trade networks by the number of trading partners

Year	Rank	Country	Outdegree Centrality C_{do}' (n_i) [%]	Country	Indegree Centrality C_{di}' (n_i) [%]
1990	1	Norway	34.9	France	32.5
(n=84)	2	Areas, nes	31.3	Spain	30.1
	3	Iceland	31.3	United Kingdom	28.9
	4	Denmark	24.1	Denmark	27.7
	5	France	21.7	FFR Germany	27.7
1995	1	Norway	42.1	USA	19.0
(n=122)	2	Areas, nes	38.0	Canada	18.2
	3	Iceland	25.6	Denmark	18.2
	4	United Kingdom	21.5	Spain	18.2
	5	USA	20.7	Portugal	14.1
2000	1	Norway	37.2	France	15.5
(n=149)	2	USA	26.4	Canada	15.5
	3	Netherlands	25.0	China	14.9
	4	United Kingdom	21.6	Spain	13.5
	5	Areas, nes	20.9	Denmark	13.5
2005	1	Norway	40.5	China	18.3
(n=154)	2	USA	30.1	Canada	17.0
	3	France	27.5	France	15.0
	4	Netherlands	24.2	Singapore	13.7
	5	Denmark	22.2	Spain	12.4
2009	1	Norway	48.0	France	18.1
(n=128)	2	USA	37.8	Nigeria	18.1
	3	France	32.3	China	16.5
	4	Spain	27.6	Spain	15.7
	5	China	26.0	Malaysia	15.0

Table 10: Degree centrality of shrimp trade networks by the number of trading partners

Year	Rank	Country	Outdegree Centrality $C_{do}'(n_i)$ [%]	Country	Indegree Centrality $C_{di}'(n_i)$ [%]
1990	1	Thailand	20.8	Spain	61.8
(n=145)	2	Denmark	20.1	France	47.9
	3	Areas, nes	18.8	USA	45.8
	4	India	18.8	Italy	42.4
	5	USA	18.8	FFR Germany	38.9
1995	1	Areas, nes	32.9	Spain	47.1
(n=156)	2	Denmark	28.4	France	41.3
	3	India	27.1	USA	34.8
	4	Thailand	26.5	Japan	34.8
	5	USA	25.8	United Kingdom	32.9
2000	1	Denmark	33.5	France	43.9
(n=174)	2	Thailand	33.5	Spain	38.2
	3	USA	30.6	USA	35.8
	4	India	28.9	Canada	30.1
	5	Indonesia	26.6	Japan	27.2
2005	1	India	45.5	Spain	37.1
(n=179)	2	China	39.3	France	32.6
	3	USA	38.2	USA	29.8
	4	Thailand	37.6	Germany	28.7
	5	Indonesia	32.6	Canada	27.5
2009	1	India	45.6	France	37.3
(n=170)	2	Viet Nam	43.8	Spain	30.2
	3	China	42.6	USA	26.0
	4	Thailand	42.6	Germany	25.4
	5	Indonesia	37.3	Belgium	24.9

Table 11: Degree centrality of salmonidae trade networks by the number of trading partners

Year	Rank	Country	Outdegree Centrality C_{do}' (n_i) [%]	Country	Indegree Centrality C_{di}' (n_i) [%]
1990	1	Areas, nes	31.5	USA	38.9
(n=109)	2	USA	29.6	France	29.6
	3	Norway	28.7	Italy	29.6
	4	Canada	27.8	FFR Germany	28.7
	5	Denmark	26.9	Spain	27.8
1995	1	Areas, nes	57.5	Saudi Arabia	22.6
(n=147)	2	USA	46.6	Canada	21.9
	3	Norway	42.5	Germany	20.5
	4	France	34.9	USA	19.2
	5	United Kingdom	30.1	France	19.2
2000	1	USA	50.8	France	21.5
(n=178)	2	Norway	41.8	Canada	20.9
	3	United Kingdom	40.7	Saudi Arabia	16.4
	4	France	39.0	USA	15.8
	5	Canada	36.7	Germany	14.7
2005	1	USA	61.0	Canada	19.2
(n=178)	2	Norway	48.6	France	18.1
	3	France	45.8	Germany	18.1
	4	Canada	44.6	USA	16.4
	5	Chile	38.4	Saudi Arabia	15.8
2009	1	USA	61.1	Germany	21.6
(n=168)	2	Norway	50.9	France	19.2
	3	Chile	45.5	Canada	18.0
	4	United Kingdom	40.7	USA	17.4
	5	France	40.1	Czech Rep.	16.2

To identify countries which play an important role as intermediaries in the fish supply chain, the betweenness centrality has been computed. Table 12 shows the top five countries by betweenness centrality for each species and year. For example, in 1990 the USA lay on 8.7 percent of the maximum possible number of geodesics of cod trading countries, it decreased to 6.4 percent in 2009. For cod and shrimps, the Asian countries China, Singapore and Vietnam, respectively, became more important as intermediaries in the last years. The main intermediaries for salmonidae are the USA, Canada and some European countries.

Table 12: Betweenness Centrality

		Cod		Shrimps		Salmonidae	
Year	Rank	Country	Between-ness Centrality $C_B'(n_i)$	Country	Between-ness Centra-lity $C_B'(n_i)$	Country	Between-ness Centrality $C_B'(n_i)$
1990	1	USA	8.7	USA	5.3	USA	9.5
	2	France	7.1	France	3.4	France	4.4
	3	Netherlands	6.5	Spain	2.8	Denmark	3.5
	4	Denmark	5.5	Japan	2.5	FFR Germany	3.4
	5	Norway	5.2	Italy	2.3	Canada	3.4
1995	1	USA	8.8	USA	8.5	USA	10.9
	2	Norway	6.2	France	5.8	France	8.5
	3	Spain	5.0	Spain	5.6	Canada	7.1
	4	Canada	4.9	United Kingdom	4.3	Belgium-Luxembourg	5.8
	5	United Kingdom	4.6	Netherlands	4.0	United Kingdom	5.8
2000	1	Norway	6.4	USA	11.9	France	12.5
	2	France	5.0	France	11.1	Canada	8.7
	3	Canada	4.3	Spain	6.5	USA	7.8
	4	USA	4.1	Canada	3.6	Norway	5.4
	5	Denmark	3.3	United Kingdom	3.4	United Kingdom	4.4
2005	1	China	7.9	USA	8.4	France	9.1
	2	Norway	6.4	Spain	7.8	USA	9.0
	3	France	6.1	France	5.7	Canada	8.9
	4	Canada	6.0	Japan	4.5	Spain	3.6
	5	Italy	5.3	China	3.0	Netherlands	3.2
2009	1	France	7.5	China	7.7	USA	7.4
	2	China	7.3	France	7.2	France	7.2
	3	USA	6.4	USA	6.4	Canada	4.6
	4	Singapore	5.1	Denmark	4.5	Italy	3.4
	5	Norway	4.9	Viet Nam	4.3	United Kingdom	3.1

5   Summary and Discussion

5.1   Summary of the results

The liberalization of the world trade, improvements in transport and cooling technologies, outsourcing of processing, and an increasing demand for fish are drivers of the growing international fish trade (FAO 2012). Trade with cod, salmonidae and shrimps increased between 1990 and 2009 by trade value and also by the number of countries involved in the trading networks of these species. The mean trade value increased for salmonidae and shrimps. Additionally, the number of trading partners has been growing steadily for the three species. The increase in the mean degree and can be interpreted as a clear increasing trend towards globalization (CASSI et al. 2009).

Cod, shrimps, and salmonidae were chosen to analyze trade networks which differ by the origin of the fish, namely fisheries and/or aquaculture. The cod trading networks differs from the networks of salmonidae and shrimps, where aquaculture became the main source of production. For salmonidae and shrimps the total trade value rose faster, the number of participating countries is higher, the average degree and the density of the network is slightly higher than it is the case for cod.

The density of the networks is relatively low with densities between 2.8 percent and 6 percent. The density reflects the degree of embeddedness of countries in a web of relationships (DE BENEDICTIS and TAJOLI 201, CASSI et al. 2009). The results show, that the worldwide fish trade network has not become more integrated and interdependent between 1990 and 2009. However, it has to be considered that the number of participating countries and trade connections increased.

The trade networks give also information on the market concentration of im- and exporting countries. In the case of the binary networks the number of trading partners increased, the median outdegree decreased, and the standard deviation of the outdegree also decreased over time. The development of these network indices may indicate that the exports of all three species concentrate on less countries, but these countries tend to export fish to an increasing number of trading partners. The increasing outdegree centralization for all three species maintains this thesis. The opposite could be detected for the importing countries. The median indegree increased over time and the standard deviation of the indegree as well as the indegree centralization decreased. The importing countries became more similar in the number of trading partners and additionally the import fish from various countries.

The betweenness centralization is relatively low for cod, salmonidae and shrimps. This indicates, that the role of countries as hubs is low and that countries directly import fish from countries where it was caught or produced. However, the development of the betweenness centrality is not the same for the three species. Comparing the values of 1990 and 2009 it could be argued that the role of intermediaries or hubs decreased for trade with cod and salmonidae and increased for shrimps. Thus the increase in trade connections for cod and salmonidae has been fairly widespread, slightly reducing the role of intermediaries.

The comparison of the top five importing or exporting countries by degree centrality shows that the main trading countries by trade value are not necessarily the same countries when the number of trade connections is analyzed. But the top five exporting and importing countries in terms of trade value or in terms of the number of trading partners do not vary by much over time. This might indicate that the consumption pattern of the consumers in the importing countries does not change much and that the production of the selected species is dominated by these exporting countries. But also climate conditions limit the existence of some fish species or the production of fish in aquaculture systems to a given area.

In the year 2009 Russia became the fifth largest importer of salmonidae, which shows that Russia might be a growing market for aquaculture products (ANDERSEN *et al.* 2009).

5.2 Discussion

Network analysis may serve as a suitable toolbox to detect changing patterns of globalization in a quantitative way (KIM and SHIN 2002). SCHWEITZER *et al.* (2009) states that network based approaches may provide a more powerful way to manage, monitor, and govern complex systems, e.g. trade networks. In addition, network analysis is able to identify structures, patterns and functional roles of agents or countries. The integration of countries in international trade is much better reflected by network parameters than by traditional trade statistics. Nevertheless, the results of network analysis are easily influenced by the number of countries selected or if cut-off values are used. To avoid such a selection bias all available trade data were used for the network analysis. Only single years have been analyzed and thus the results may be biased by sporadic events, e.g. disease outbreak in salmon farming. Aggregation of some years could lead to the detection of stable patterns which are not influenced by sporadic events (CASSI *et al.* 2009).

A previous study on the quality of fish trade data showed, that the data are predominantly inaccurate and that the consistency of the data gets worse the more disaggregated the product classifications are (GUETTLER 2013, Chapter 3 in this dissertation). This study used disaggregated trade data to analyze species specific differences in the trade networks. Biased results caused by data quality may also appear in this study. In light of this, the results should be interpreted with care.

The analysis of fish trade networks over a longer time horizon would have been of interest but species specific trade data are only available since the introduction of SITC Rev. 3 in 1988. The analysis of trade data on a higher aggregated commodity level would have been possible for a longer period but this clearly limits the explanatory power for the detailed species.

In a further step it could be useful to identify blocks of countries which have similar characteristics in the network and to identify countries which define the core of the trade network.

5.3 Concluding remarks

This study analyzed the properties of international fish trade networks and their evolution over time. Data on commodity specific trade flows, namely cod, salmonidae and shrimps were used. Topological properties of the networks have been calculated and compared with each other. The results indicate a globalization process of fish trade and differences between capture- and aquaculture based species. The increasing aquaculture production of shrimps in Asia and for salmon in Norway and Chile made these countries also to the main exporters of these species. The main importers are Japan, USA and European countries. Japan has traditionally a high consumption of seafood. In Europe overfishing and increasing demand for fish led to increasing imports of fish and thus a dependency on other fish producing and exporting countries emerged.

Aquaculture production is predicted to increase and will play an important role to satisfy future demand of fish (FAO 2012). Thus it is interesting to monitor the effects on trade and trade networks when aquaculture becomes more and more important.

References

Andersen, T.B., Lien, K., Tveterås, R. and Tveterås, S. (2009): The Russian seafood revolution: shifting consumption towards aquaculture products. Aquaculture Economics & Management 13(3): 191-212.

Asche, F. (1997): Trade disputes and productivity gains: The curse of farmed salmon production? Marine Resource Economics 12(1): 67-73.

Asche, F., Guttormsen, A.G. and Tveterås, R. (1999): Environmental problems, productivity and innovations in Norwegian salmon aquaculture. Aquaculture Economics and Management 3(1): 19-29

Borgatti, S.P., Everett, M.G. and Freeman, L.C. (2002): Ucinet for Windows: Software for Social Network Analysis. Harvard, MA: Analytic Technologies.

Cassi, L., Morrison, A. and Ter Wal, A. (2009): The evolution of knowledge and trade networks in the global wine sector: a longitudinal study using social network analysis. Papers in Evolutionary Economic Geography, # 09.09, Utrecht University. <http://econ.geo.uu.nl/peeg/peeg.html>

De Benedictis, L., and Tajoli, L. (2009): Comparing international sectoral trade networks. <http://works.bepress.com/cgi/viewcontent.cgi?article=1012&context=luc a_de_benedictis>

De Benedictis, L., and Tajoli, L. (2011): The world trade network. The World Economy 34(8): 1417-1454.

FAO (2011): FAO Fisheries and Aquaculture Department, Statistics and Information Service, FishStatJ: Universal software for fishery statistical time series. Copyright 2011.

<http://www.fao.org/fishery/statistics/software/fishstatj/en>

FAO (2012): The state of world fisheries and aquaculture 2012. Food and Agriculture Organization of the United Nations, Rome.

FAO/WHO (2011): Report of the joint FAO/WHO expert consultation on the risks and benefits of fish consumption. Food and Agriculture Organization of the United Nations, Rome, World Health Organization, Geneva.

Freeman, L.C. (1977): A set of measures of centrality based on betweenness. Sociometry, 40(1): 35-41.

Freeman, L.C. (1979): Centrality in social networks: Conceptual clarification. Social Networks 1: 215-239.

Gould, R.V. (1987): Measures of betweenness in non-symmetric networks. Social Networks 9: 277-282.

Guettler, S. (2013): Testing the quality of international fish trade data. In: In: Guettler, S.: Studies on the market assessment of aquaculture research by a small producer country, Dissertation, Kiel University, 17-66. <http://eldiss.uni-kiel.de/macau/receive/dissertation_diss_00010683>

Kim, S. and Shin, E.H. (2002): A longitudinal analysis of globalization and regionalization in international trade: a social network approach. Social Forces 81(2): 445-468.

Mozzafarin, D. and Rimm, E.B. (2006): Fish intake, contaminants, and human health: evaluating the risks and the benefits. Journal of the American Medical Association 296(15): 1885-1899.

Schweitzer, F., Fagiolo, G., Sornette, D.,. Vega-Redondo, F, Vespignani, A. and White, D.R. (2009): Economic networks: the new challenges. Science 325: 422-425.

Scott, J. (2000): Social network analysis – a handbook. 2nd ed., Sage Publications, London, Thousand Oaks, New Delhi.

UN Comtrade (2011): United Nations Commodity Trade Statistics Database. http://comtrade.un.org/db/default.aspx

Wasserman, S., and K. Faust (1994): Social Network Analysis: Methods and Applications. Cambridge University Press, Cambridge.

5. The shape of future aquaculture R&D – results of a Delphi study

Susanne Stricker, Stefan Guettler, Carsten Schulz, Rolf A.E. Mueller

Published in: Aquaculture Europe 34(2), 2009, pp. 18-20.

5 The shape of future aquaculture R&D – Results of a Delphi study

Introduction

The future direction that research and development (R&D) will take in any one field is always shrouded by thick mist. The mist is, however, never impenetrable. In the summer 2008 we launched a worldwide online Delphi study to penetrate the mist that shrouds the future of aquaculture R&D. The purpose of the project was to contribute towards focussing better the nascent aquaculture research program of the Faculty of Agriculture and Nutritional Science of the University at Kiel, Germany.

The Delphi method is a widely-used and well-accepted method for casting light on future developments in a certain domain by systematically and repeatedly interrogating experts for this domain and by synthesizing experts' opinions (Linstone and Turoff 1975). In some sense, the Delphi method can be characterized as a form of collaborative qualitative forecasting by geographically dispersed domain experts. We employ this method to ascertain the likely future directions of aquaculture R&D in developed, high-income countries. This study comprises three survey rounds. Goal of the initial round was to assess the current situation and anticipated future developments of research in aquaculture in the long run (until the year 2020), as seen by aquaculture experts. The second and third round focussed on finding consensus on specific fields as well as to discuss fields identified as promising for R&D investments in more detail. Our study focuses on aquaculture of finfish in advanced economies. Other species than fish, such as crustaceans, molluscs, aquatic plants, etc. and world regions, other than the high-income countries, such as China where aquaculture is particularly important, were outside the scope of our study.

Based on a bibliometric study of co-author networks in aquaculture and fisheries research (Seidel-Lass 2009) we developed a list of some 1,300 email addresses. The addresses were contacted by email and invited to participate in the Delphi survey which we conducted on the web. After two reminder emails 272 (21 percent of the addressees contacted) aquaculture researchers participated in the first round. This response rate is remarkably high compared to other online surveys, where response rates below one percent have been reported.

Issues and Delphi rounds

The Delphi study was preceded by a survey of R&D issues in the aquaculture R&D literature (Guettler 2008). This set of issues was augmented and focused in consultations with a small number of aquaculture R&D experts. Based on this list of R&D issues we developed the questionnaire for the first Delphi round. This questionnaire comprised 45 questions which were organized into the following sections:

1. About the respondents
2. General questions
3. Fish breeding and reproduction
4. Fish husbandry and water management
5. Fish health
6. Fish nutrition
7. Marketing and quality management

The questionnaire of the second round was much shorter, comprising only 14 questions, while the third round questionnaire contained 10 questions.

The following sections present the key results of the Delphi study.

The respondents

The aquaculture experts participating in the survey were between 26 and 78 years old, their average age was 48 years. The vast majority of aquaculture experts hold a PhD degree (73 percent) and their professional experience in aquaculture spans a 19 year period, on average. While three fourths of the responding experts have between 11 and 25 years of experience in aquaculture R&D, half have 20 years or more of aquaculture research experience. Nearly half of the respondents are employed by Universities and nearly every fifth respondent works for a governmental agency. Most of the respondents are professors or senior researchers and focus on applied or basic research. The respondents currently live in all parts of the world, most in Europe, northern America, Norway, and the United Kingdom.

General aspects of aquaculture R&D

Aquaculture experts agree on two points: (i) aquaculture research in general has achieved much and will continue to do so in the future, and (ii) aquaculture research achievements will have a very strong impact on the productivity of aquaculture as well as on the quality of fish produced.

R&D rarely progresses in lock-step on all research fronts. For past and current research achievements fish nutrition was rated highest by the respondents, followed by breeding, reproduction, fish husbandry, fish health, water management, quality management, fish marketing and finally organic aquaculture (Tab. 1). The ranking was, however, significantly modified with regard to future expected research achievements. In the future, fish health will come first, followed by fish nutrition, quality management, and water management.

Table 1: Average scores of expert rating of past and current achievements as well as future development of aquaculture research by areas (n=272)

Research area	Average score		Rank past and current	Rank 2020
	Past and current*	Development until 2020**		
Fish nutrition	4.01	3.95	1	2
Breeding	3.91	3.74	2	7
Reproduction	3.91	3.61	3	8
Fish husbandry	3.75	3.59	4	9
Fish health	3.71	4.06	5	1
Water management	3.58	3.93	6	3
Quality management	3.22	3.93	7	3
Fish marketing	3.00	3.81	8	6
Organic aquaculture	2.51	3.87	9	5

* scale from 1 = poor to 5 = very substantial
** scale from 1 = much less to 5 = much more

Will the hotspots of aquaculture R&D shift in the future? Norway was by far rated as the current and future leading aquaculture research nation. Spain and the USA are expected to become much stronger players in aquaculture research. Germany and Italy were rated very low and our experts expect these nations to continue to linger at the bottom of the aquaculture R&D charts. Interestingly, the mean ratings of the future development of the nations' strengths in aquaculture research lie very much closer together than the average ratings of current strengths (Tab. 2).

Table 2: Advanced economies and their current strength in aquaculture R&D (n=272)

Country	Mean current strength*	Anticipated future strength**	Rank current	Rank future
Norway	4.64	3.75	1	1
Israel	3.68	3.55	2	5
United Kingdom	3.63	3.39	3	8
Canada	3.57	3.56	4	4
USA	3.50	3.58	5	2
France	3.45	3.40	6	7
Denmark	3.34	3.33	7	10
Spain	3.28	3.58	8	2
The Netherlands	3.26	3.35	9	9
Greece	2.92	3.46	10	6
Germany	2.76	3.31	11	11
Italy	2.74	3.27	12	12

* scale from 1 = very weak to 5 = very strong
**scale from 1 = much less to 5 = much more

R&D on organic aquaculture was considered to have produced the least results so far. Moreover, most experts agreed that organic aquaculture is generally overrated; nevertheless, R&D spending on organic aquaculture is expected to increase considerably.

Fish breeding and reproduction

There was consensus among almost all (98%) aquaculture experts that developing breeding programs comparable to livestock breeding programs would be useful or perhaps even very useful for fish species. Among the many fish species suitable for improvement through systematic breeding, European

Seabass, Gilthead Seabream, and Turbot were, in this order, rated as particularly promising.

With regard to specific research areas concerned with fish breeding and reproduction, there was consensus that achievements were highest in conventional selective breeding, followed by chromosome set and sex manipulation, and crossbreeding. For the future our experts predict that marker based selective fish breeding will develop into a highly productive research area.

Fish husbandry and water management

How attractive for R&D are recirculating systems, cage systems, pond, and flow-through systems? Our experts agree that research expenditures on recirculating systems will increase most substantially, followed by R&D investments into cage systems. Very few, however, expect that more R&D money will be flowing into ponds or towards flow-through systems.

Because results of the first two Delphi rounds highlighted the high importance of R&D on recirculation systems, we decided to probe a bit deeper and ask our respondents to rate the importance of specific research issues related to recirculation systems. Energy efficiency, nutrient discharge, and biological clarification systems are considered to be the most promising research areas in connection with recirculation systems. Research on the material and shape of the fish rearing unit were clearly rated lowest and the potential of this research was rated somewhere between "little" and "some" potential.

Research on fish husbandry and water management is not exhausted by research on specific water systems. Other areas which are expected by many to attract considerable attention and funding over the next decade are integrated multitrophic aquaculture (IMTA) as well as the environmental impact of aquaculture.

The environment is vast and aquaculture's impacts on the environment can be many. We therefore probed for research on specific impacts, distinguishing between carbon, carbon dioxide, nitrogen, and phosphorus. Our respondents believe that R&D on aquaculture's nitrogen and phosphorous related environmental impacts has, in the past, achieved more results than research on impacts related to carbon. This is, however, expected to change. In the future, research on all four impacts is expected to be on a similar level of achievement.

Fish health

Within the research fields concerning fish health, aquaculture experts expect a dynamic change in research fields with the greatest achievements. While research on bacteria and parasites were rated highest for their past and current research achievements, the future will be in developing therapeutics and vaccines, disease gene mapping, and early identification systems.

Fish nutrition

The contribution of aquaculture toward achieving sustainable food security for a growing world population will be questioned as long as aquaculture fish are fed with processed captured fish. There is hope that this blemish will soon be removed from aquaculture. Our experts think that R&D on vegetative resources and derivates is very likely to yield until 2020 alternatives for fish meal and fish oil. Among the many potential plant resources that might provide the alternative feed compounds, legumes are considered to be the most promising source for alternative for fish meal whereas oleiferous fruits are considered to be the most important sources for fish oil substitutes. Few will be surprised to learn that our experts think that R&D on finding substitutes derived from potatoes is unpromising.

The hope to escape the dependence on fish caught in the wild is shared by many. Two thirds of the aquaculture experts participating in the survey think it will be possible to achieve feed conversion ratios below one when feeding carnivorous species. Three quarters of the respondents think feed conversion ratios smaller than one will be achieved by 2015. There is, however, considerable uncertainty as to the period when feed conversion ratios break through the barrier of 1: a standard deviation of 8.8 years suggests that breaking the barrier is as likely to happen tomorrow as in the year 2024.

Marketing, quality management and economics of aquaculture

Most aquaculture experts think that current advances in research on fish and fish products is less or even much less advanced than research on meat and meat products. However, this is the one result on which our experts achieved no consensus.

Quality management was rated little better than marketing research, even though the average of all rating indicates that research on aquaculture fish quality management has not reached the same status as quality management research on meat and meat products

Even though there is dissent among the experts about current achievements, there is consensus that research on fish marketing and fish quality management will substantially increase until the year 2020.

Aquaculture experts think that past and current achievements on aquaculture farm business management were highest, while they assume that the development of achievements will be highest in traceability and supply-chain-management.

Conclusions

Our Delphi study showed that aquaculture experts are commonly convinced by the high R&D achievements in the past and the future in order to increase productivity of aquaculture systems. Especially, improvements in the field of fish nutrition were identified as major strength in the past and fish health aspects will be focussed primarily in future R&D, followed by fish nutrition. Aquaculture experts are aware that efficient resource utilization in terms of e.g. feed conversion, energy utilization, nutrient discharge, water reuse technologies or pathogen treatment will be major challenges for upcoming research activities.

Highlighted priorities identified by our Delphi study among aquaculture experts should not only forecast R&D activities, but also should help funding agencies and decision maker to identify relevant areas of interests.

References

Guettler, S. (2008): Forschung und Entwicklung in der Aquakultur – ein Überblick über Arbeitsgebiete und offene Fragen. I&I Working Paper, <http://www.agric-econ.uni-kiel.de/Abteilungen/II/veroeffentlichungen.shtml>

Linstone, H.A. and Turoff, M. (1975): The Delphi Method – Techniques and Applications. Addison-Wesley, London.

Seidel-Lass, L. (2009): Networks in International Aquaculture Research: a Bibliometric Analysis. Cuvillier Verlag, Göttingen.

6. Demand for fish in Germany

6 Demand for fish in Germany

1. Introduction

Fish is known as healthy food and nutritionists recommend that fish is eaten twice a week (FAO/WHO 2011, Mozzafarin and Rimm 2006, Perk et al. 2012). In Germany the average annual consumption of fish was about 15.3 kg per capita in the year 2007, significantly below the average annual per capita consumption in the EU of 23.3 kg (EU 2012). Moreover, per capita consumption of fish stagnates in Germany since the year 2000, whilst global per capita consumption has increased considerably (FAOSTAT 2012). One reason for the stagnating per capita consumption might be the development of fish prices in the past. The FAO fish price index indicates that real global fish prices have increased by more than 55 percent between 2002-2004 and 2012 (FAO 2012). In comparison real fish prices in Germany increased only moderately by 23 percent between 2005 and 2011 (Destatis 2011).

The consumer reaction to price and income changes is usually measured by elasticities of demand. The knowledge of such elasticities is useful for guiding policy measures. However, not much is known about current demand for fish in Germany because it has not been studied for quite some time. The studies by Ryll (1984) and Sommer (1985) are the only studies which are explicitly concerned with this issue. The aim of this study is to fill this gap by estimating price and expenditure elasticities for fish demand in Germany. To this end, cross sectional household data from the 2003 German income and consumption survey (EVS 2003) are used to estimate price elasticities and expenditure elasticities of demand for fish, several fish products, as well as for other food commodities. The EVS 2003 data represent 98 percent of German households and allow the estimation of a demand system. A standard estimation procedure is applied to the EVS data to gain information on own-price and cross-price elasticities, as well as expenditure elasticities. Demand is analyzed in three stages. In the first stage a linear expenditure system is used to estimate the expenditures for food. In the second and third stage a quadratic almost ideal demand system (QUAIDS) is applied to analyze the demand for several fish and food products.

The estimation of a demand system requires information on prices and quantities demanded of each commodity. Unfortunately, cross sectional data are not always perfect. Price information is often not directly available and not every consumer consumes each good which leads to zero observations in the data set. To overcome the problem of missing price information, a method based on Cox and Wohlgenant (1986) may be applied. This method uses a hedonic price function to estimate commodity prices. The data set may also contain zero observations, e.g. zero consumption of a commodity. Estimating a demand system without these zero observations may lead to biased results. To avoid this selectivity bias a two-step estimation procedure may be applied (Shonkwiler and Yen 1999). To control for the effect of these two methods on the estimation results, a sensitivity analysis is conducted additionally.

The paper proceeds as follows. First, the literature on fish demand in Germany and some other countries is reviewed. Section 3 gives a short overview on the cross sectional household data. In section 4 the model and estimation procedure is specified before the results on elasticities, including a sensitivity analysis are presented in section 5. The paper ends with a discussion and a summary.

2. Review of the literature on fish demand

2.1 Review of the literature on fish demand in Germany

In the last 30 years only two publications explicitly scrutinized the demand for fish in Germany, namely Ryll (1984) and Sommer (1985). Some other papers analyzed the demand for fish among other products and estimated demand elasticities. This section gives an overview on the results of these studies, which are also summarized in Table 1 at the end of this section.

Ryll (1984) analyzed demand for several fish products between 1961 and 1980. Simple regression analysis was used to determine the influence of several product prices on the quantity demanded by households and fish products. Results for income elasticity, own price elasticity and cross price elasticity vary by product and time. The average income elasticity varies from -0.64 for fresh fish to 3.49 for crustaceans and mollusks. The own-price elasticity ranged from -0.23 for smoked herring, sprats, and other smoked fish, up to -6.29 for canned tuna. Ryll identified poultry meat as the main substitute for fish with a cross-price elasticity of 0.34.

For the period from 1965 to 1981 Sommer (1985) investigated the demand for fish for four-person-households based on their incomes and expenditures for fish. Although the focus of this study lies on the detection of seasonal demand patterns, Sommer also computed demand elasticities. Based on linear regression the own price elasticities for fresh or frozen fish vary between -1.28 for households with an average income to -0.51 for households with higher income. The income elasticity was lowest for households with an average income and the period from 1965 to 1974 with a value of 0.55 while the highest income elasticity with a value of 1.32 was detected for the same household type between 1978 and 1981.

Michalek and Keyzer (1992) used macroeconomic time series data and applied a two-stage LES-AIDS model to analyze the demand for ten food products in eight European countries between 1970 and 1985. The income elasticity for fish was 0.03 in 1970 and 0.02 in 1985 in Germany and significantly lower than in other countries, e.g. Ireland (2.05 in 1970 and 1.97 in 1985), or other food products, e.g. meat (0.29 in 1970 and 0.24 in 1985)[4]. The Marshallian[5] or uncompensated own-price elasticity, which includes the income effect and the substitution effect, was -0.05 in 1970 and -0.06 in 1985 in Germany. The Hicksian[6] or compensated own-price elasticities, which include the substitution effect and exclude the income effect, were not calculated by Michalek and Keyzer (1992).

Henning and Michalek (1992) estimated a nested three-stage LES-LES-AIDS system for German food demand. The income elasticity for fish increased from 0.30 to 0.41 between 1970 and 1985 while the Marshallian own-price elasticity changed from -0.13 in 1970 to -0.70 in 1985.

---

[4] Chavas (1983) names technological adoption, shifts in consumer preferences and institutional changes as sources for varying demand elasticities over time. In the case of meat demand, consumers concerns on fat and cholesterol may have led to a shift of meat preferences.

[5] Marshallian demand functions depend on prices and income.

[6] Hicksian demand functions depend on prices and utility.

Schons` (1993) time series analysis with OLS regression between 1965 and 1988 leads to a negative income elasticity of -0.45 for fish in Germany.

Pawlik (1993) analyzed the demand for frozen food between 1975 and 1989 in Germany with a linear regression. The results show that the income elasticity for frozen fish fillets decreases from 1.10 in 1975 to 0.74 in 1989 and that the own-price elasticity shifts from -1.13 to -0.75 in the same period.

Wildner and von Cramon-Taubadel (2000) estimated a two-stage LA/AIDS for four-person households with higher income (monthly income of 6,800 DM to 9,050 DM, approximately 3,476 € to 4,627 €), where on the second stage the demand for fish and fish fillets is analyzed between 1966 and 1997. The Marshallian own-price elasticity is estimated to be -0.46 while the main substitute is poultry with a cross-price elasticity of 0.5. Additionally, beef and veal is found to be a complementary good with a cross-price elasticity of -1.48.

Thiele (2001) applied a LES to estimate expenditure and price elasticities of food commodities using German cross-sectional household data (Einkommens- und Verbrauchsstichprobe, EVS) of the year 1993. The own-price elasticity of fish is -0.72 while the expenditure elasticity for fish is 0.70. Thiele also shows that demand for fish is elastic for low-income households with a price elasticity of -1.04, while it is inelastic for high-income households with a price elasticity of -0.60. The same holds for young households, where the main income recipient is aged below 30 years (-1.18) and old households, where the main income recipient is aged 60 years and older (-0.33) or 1-person households (-1.29) and couples with two children (-0.80), respectively. The expenditure elasticity for fish is smallest for young households (0.34) while the highest expenditure elasticity was found for old households (0.99).

Wildner (2001a) estimates price and expenditure elasticities for food in the former West German states with a LA/AIDS system. Monthly data between 1966 and 1997 of four-person households with higher income (monthly income of 6,800 DM to 9,050 DM, approximately 3,476 € to 4,627 €) were used to calculate elasticities. The Marshallian expenditure elasticity for fish and fish fillets was 0.54 in 1970 and increased to 0.72 in 1995. The Marshallian own-price elasticity was -0.94 and -0.96 in 1970 and 1995, respectively. The Hicksian own-price elasticity is -0.93 in both years. Wildner also estimates cross-price elasticities. The highest cross-price elasticity is found for sausages: 0.22 in 1970 and 0.13 in 1995 for Marshallian elasticities, and 0.50 and 0.55 for Hicksian elasticities.

The study of Wildner (2001b) is an extension of Wildner (2001a) to a total of three different household types. For 2-person households with an income between 1,750 DM and 2,650 DM (894 € to 1,355 €) the Marshallian expenditure elasticity for fish and fish fillets increased from 0.21 in 1970 to 0.54 in 1995, while the own-price elasticity varied from -0.46 to -0.69 in the same period. Pork was identified as the main substitute with a cross price elasticity of 0.24 in 1970 and 0.16 in 1995. The Hicksian own-price elasticity was -0.45 in 1970 and -0.65 in 1995 and the highest cross-price elasticity with a value of 0.4 were found for sausages. For a couple with two children and an income between 3,850 DM and 5,850 DM (1,968 € to 2,991 €) the Marshallian expenditure elasticity was 0.62 in 1970 and 0.72 in 1995 and the own-price elasticity of fish and fish fillets was elastic with values of -1.13 and -1.10, respectively. For this household type, poultry was found to be a compliment for fish with significant cross-price elasticities of -0.23 and -0.18. The Hicksian own-price elasticity changes slightly from -1.12 in 1970 to -1.08 in 1995.

Hoffmann (2003) used German cross-sectional household data (EVS) from 1998 to estimate a 3-stage LES. The income elasticity for fish decreased from 0.28 in the first quarter of 1998 to 0.19 in the second, to 0.08 in the third and to 0.04 in the fourth quarter. The expenditure elasticity for fish is quantified with 0.76. Price elasticities are not calculated by Hoffmann (2003).

Previous studies, which cover the period between 1961 and 1998, show that the estimated price elasticities vary much, depending on the data used, period regarded or method applied. The demand for fish seems to be inelastic in general with values mainly between -0.4 and -0.9. Poultry seems to be the main substitute for fish with cross-price elasticities between 0.3 and 0.5. But, poultry is sometimes also identified as a complement to fish. The income or expenditure elasticities are positive but mostly below 1, indicating that fish is seen as a necessity by the German households. Table 1 summarizes the literature overview on fish demand analysis in Germany.

Table 1: Selected elasticities for German fish demand

Source	Data source	Method	Period	Elasticity				identified substitute/ complement
				Income	Expenditure	Own-price	Cross-price	
Ryll (1984)	yearly macro-economic data	Regression analysis	1961-1980	0.80		-1.37	0.34 poultry	
Sommer (1985)	individual business accounts [3]	Regression analysis	1965-1981	0.55 to 1.32		-0.52 to -1.28		
Schons (1993)	yearly macro-economic data	Regression analysis	1965-1988	-0.45				
Pawlik (1993)	yearly macro-economic data	Regression analysis	1975-1985	0.74 to 1.1		-0.75 to -1.13		
Thiele (2001)	cross-sectional data	LES	1993		0.70	-0.72		
Hoffmann (2003)	cross-sectional data	LES	1998	0.04 to 0.28	0.76			
Michalek, Keyzer (1992)	time series data	LES AIDS	1970-1985	0.02 to 0.03		-0.05 to -0.06 [1]		
Henning, Michalek (1992)	yearly macro-economic data	LES LES AIDS	1970-1985	0.30 to 0.41		-0.13 to -0.70 [1]		
Wildner, von Cramon-Taubadel (2000)	individual business accounts [3]	LA/AIDS	1966-1997		-0.127	-0.463 [1]	0.495 poultry	
Wildner (2001a)	individual business accounts [3]	LA/AIDS	1966-1997		0.54 to 0.72	-0.94 to -0.96 [1] / -0.93 [2]	0.13 to 0.22 sausages / 0.50 to 0.55 sausages	
Wildner (2001b)	individual business accounts [3]	LA/AIDS	1966-1997		0.21 to 0.72	-0.46 to -1.13 [1] / -0.45 to -1.12 [2]	-0.06 to -0.23 poultry / 0.22 to 0.72 sausages	

1) Marshallian (uncompensated) price elasticity
2) Hicksian (compensated) price elasticity
3) Laufende Wirtschaftsrechnung

2.2 Review of the literature on fish demand in other countries with similar food habits and income level

A vast amount of fish demand studies exist, which cannot be reviewed here in detail. Therefore three literature review studies are briefly summarized and three other studies on fish demand in France and Canada are presented.

Andreyeva et al. (2010) reviewed 18 US-based studies on fish demand. The mean own-price elasticity of fish is -0.5 in the USA. The range of elasticity estimates spans from -0.05 to -1.41 and the 95 percent confidence interval is estimated between -0.3 and -0.69.

Gallets' (2009) meta-analysis of 168 studies representing over 1,000 own-price elasticities showed a median elasticity for fish of -0.8. The median own-price elasticity for the USA is -1.08 and -0.68 for the United Kingdom.

The review of demand studies of detailed fish species in developed countries by Asche et al. (2005) finds that demand for fish is price elastic in most markets. The main substitutes for seafood products tend to be other seafood products while the degree of substitution between seafood and meat is substantially less.

Bjørndal et al. (1992) analyzed the French market for salmon in the 1980ies. The short run own-price elasticity of salmon demand in France is estimated to be -1.06 and the long run elasticity is about -1.3.

Lambert et al. (2006) analyzed the demand for fish and meat in Canada and find a Marshallian own-price elasticity of fish between -0.48 and -0.77 and a Hicksian own-price elasticity between -0.4 and -0.71.

Salvanes and De Voretz (1997) estimate a compensated and uncompensated own price elasticity of fish of about -0.88 for Canada. The own-price elasticity of fresh, cured and canned fish is nearly one-elastic. The expenditure elasticity of all analyzed fish species is about 1.

This short review of international demand studies indicates that the demand for fish in developed, high-income countries seems to be inelastic. But, the demand analysis of detailed fish species shows, that demand seems to be elastic when disaggregated fish commodities are analyzed.

3. Cross-sectional household data

Cross-sectional household data of the year 2003 serve as data basis for this study. The German Federal Statistical Office (Destatis) conducted an income and consumption survey in the year 2003 (EVS 2003). The survey was executed throughout the whole year. A sub-sample of nearly 12,000 out of more than 53,000 private households in Germany in the total sample recorded their expenditures on food, beverages and tobacco for one month. These households represent 98 percent of the German households (Destatis 2005).

The mean household in the sub-sample of 12,000 households, which recorded their food expenditures, has a monthly net income of nearly 3,500 € and the average household size is 2.42 persons.

In 2003, private households used 10.1 percent of their consumption expenditures for food. Monthly expenditures for food were 197 € per household, of which the largest part with 24 percent were expenditures for meat and meat products, followed by bread and cereals (18.6 percent), dairy products (16.2 percent), vegetables and potatoes (11.9 percent), and fruits (10.1 percent). Only 3.4 percent of the monthly food expenditures or 6.74 € were spent on fish and fish products (Czajka and Kott 2006).

Table 2 shows the expenditures for food and fish by income class. The expenditures for food increase with rising income, but the share of fish expenditures on total food expenditures is nearly constant over the income groups with a share between 3.1 and 3.7 percent. In contrast, the likelihood that fish is consumed increases with the household income: Only 61 percent of the households with a monthly net income below 900 € consume fish, while this share increases with rising income to 81 percent of the households in the income class from 5,000 to 18,000 € per month.

Table 2: Monthly expenditures for food and fish and share of fish consuming households by income

	all house-holds	monthly net household income from... to below... Euro							
		<900	900 - 1300	1300 - 1500	1500 - 2000	2000 - 2600	2600 - 3600	3600 - 5000	5000 - 18000
expenditures for food [€]	196.56	94.64	116.61	130.08	151.34	186.38	223.82	259.17	292.93
expenditures for fish [€]	6.74	3.10	3.98	4.59	5.57	6.51	7.58	8.14	10.58
expenditures for fish as share of total food expenditures [%]	3.4	3.3	3.4	3.5	3.7	3.5	3.4	3.1	3.6
share of fish consuming households [%]	76.3	61.1	65.3	67.0	71.6	77.0	79.1	79.4	81.1

Source: Czajka and Kott 2006, own calculations.

4. Methodology for demand analysis and estimation procedure

The theory of demand is the foundation of the econometric estimation for demand systems. An overview of demand theory is given by Deaton and Muellbauer (1980b), Deaton (1986), and more recently by Barnett and Serletis (2008) and Okrent and Alston (2011). The latter two studies also review the literature on demand systems and their estimation.

Two basic approaches exist for the estimation of demand elasticities: single equation models and demand systems. In single equation models, demand for a commodity is a direct function of income and prices (Intriligator 1978, Lau 1986, Ecker and Qaim 2011). These single equation models are easy to estimate but they may be inconsistent with demand theory (Lau 1986, Ecker and Qaim 2011, Okrent and Alston 2011). Demand systems consist of several functions which are estimated simultaneously. Demand systems are consistent with demand theory and are able to reflect interdependencies between several products (Okrent and Alston 2011).

4.1 Multistage budgeting

The EVS 2003 data-set contains information on more than 200 commodities. The estimation of a full demand system is not practical for such data, as the number of own- and cross-price elasticities increases with the square of the number of commodities (Gao et al. 1996, Edgerton 1997). To overcome this difficulty a multistage budgeting framework is used in this study to model the fish demand of households in Germany. Deaton and Muellbauer (1980b) and Edgerton (1997) show that multistage budgeting is only consistent with demand theory, if preferences are weakly separable and the price index is a good approximation of the true cost of living index. Under these conditions the usual assumption in the multistage budgeting framework is that the consumer decides on the information of price indices on the expenditure allocation to commodity groups. In the next stage, the expenditure allocation within the commodity groups is performed independently of the other commodity groups (Edgerton 1997). The multistage-budgeting approach is for example applied by Blundell et al. (1993), Gao et al. (1996), Edgerton (1997), Dey (2000), Dey et al. (2005, 2008, 2011), Kumar et al. (2005) and Pan et al. (2008).

In this study it is assumed that in the first stage a household allocates its income or expenditures to food commodities and non-food commodities. In the second stage the expenditures for food are allocated to specific food commodity groups, i.e. fish, meat, dairy products, fruits and vegetables. In the third stage the expenditures for fish are disaggregated into more specific fish products, in particular fresh or frozen fish and fish fillets. The estimated food expenditures from the first stage are then used in the second stage. For the remaining stages two and three a quadratic almost ideal demand system (QUAIDS) developed by Banks et al. (1997) is applied to gain detailed information on expenditure elasticities, own-price elasticities and cross-price elasticities of the commodities.

4.2 Linear food expenditures

In the first stage, the expenditures for food depend on the Stone price index for food pf and a price index for non-food commodities and the monthly income of the household. The Stone price index for food is computed as the geometric mean of the food prices $pf = \sum w_j p_j$, where p_j is the price of commodity j and w_j is the share of commodity j in total food expenditure. Due to data constraints, the difference between income and food expenditures is used as a proxy variable for the price index for non-food commodities. Further household characteristics are used to estimate the expenditures for food, namely the number of persons living in the household, age and education of the main income recipient of the household, region, and whether or not the household is a single male household.

The food expenditure function is specified as follows:

(1) $\ln expf_h = \beta_0 + \beta_1 \ln pf_h + \beta_2 \ln pnf_h + \beta_3 \ln y + \sum_{i \in z} \beta_i \, hc_{ih} + e_h$, with

   $expf$ expenditures for food

   pf Stone price index for food

   pnf price index for non-food commodities

   y monthly household income

   hc vector of household characteristics

   e error term

Equation (1) can be derived from a household utility function which is maximized subject to a budget constraint. The estimate of equation (1) is constrained by the homogeneity restriction of degree zero in prices and income, which is specified by (2) (Intriligator 1978, Blundell et al. 1993, Dey et al. 2005):

(2) $\beta_1 + \beta_2 + \beta_3 = 0$

4.3 The quadratic almost ideal demand system (QUAIDS)

The foundation of the QUAIDS is the almost ideal demand system (AIDS) by Deaton and Muellbauer (1980a). The budget share equation of the AIDS is derived from utility theory by applying Shephard's lemma to an expenditure function and transforming the resulting Hicksian demand functions to Marshallian demand functions. The AIDS has the property to satisfy the axioms of consumer choice (consumer preferences are complete, continuous, and transitive) exactly and it can be used to test the restrictions of homogeneity and symmetry (Deaton and Muellbauer 1980a). Banks et al. (1997) improved the AIDS by adding a quadratic expenditure term to reflect the empirically fact that Engel curves are not always linear (Lewbel 1991, Blundell et al. 1993, Banks et al. 1997). Schmitz (2007), who also used the EVS 2003 data, showed that Engel-curves are non-linear for fish and meat and other food groups. Thus the QUAIDS model is therefore deemed appropriate for this study.

Banks et al. (1997) assume an indirect utility function of the form

(3) $\ln V = \left\{ \left[\frac{\ln m - \ln a(p)}{b(p)} \right]^{-1} + \lambda(p) \right\}^{-1}$,

where ln V is the indirect utility function, which depends on the expenditures m and functions of the price vector $\boldsymbol{p}$. The translog price index ln $a(\boldsymbol{p})$ is definded as

(4) $\ln a(\boldsymbol{p}) = \alpha_0 + \sum_{i=1}^{K} \alpha_i \ln p_i + \frac{1}{2} \sum_{i=1}^{K} \sum_{j=1}^{K} \gamma_{ij} \ln p_i \ln p_j,$

where $i = 1, ..., K$ denotes the number of goods

The Cobb-Douglas price aggregator $b(\boldsymbol{p})$ is:

(5) $b(\boldsymbol{p}) = \prod_{i=1}^{K} p_i^{\beta_i}$

The specific functional form of $\lambda(\boldsymbol{p})$ is:

(6) $\lambda(\boldsymbol{p}) = \sum_{i=1}^{K} \lambda_i \ln p_i,$

where $\sum_{i=1}^{K} \lambda_i = 0$.

Applying Roy´s identity to (3), the budget share equation for each commodity group is presented by (7):

(7) $w_i = \alpha_i + \sum_{j=1}^{K} \gamma_{ij} ln P_j + \beta_i ln \left\{ \frac{m}{a(p)} \right\} + \frac{\lambda_i}{b(p)} \left[ln \left\{ \frac{m}{a(p)} \right\} \right]^2,$

where w_i is the budget share of each household for commodity i, m indicates the household income and P_j are the prices of the commodities.

The parameters α_i, β_i, γ_{ij}, and λ_i have to be estimated, where β_i measures the effect of a real income change to the change in budget share of commodity i, γ_{ij} measures the effect of a price change of commodity j on the budget share of i. For theoretical consistency equation (3) is estimated under additivity (8), homogeneity (9), and symmetry restrictions (10):

(8) $\sum_{i=1}^{K} \alpha_i = 1 ; \sum_{i=1}^{K} \beta_i = \sum_{i=1}^{K} \lambda_i = \sum_{i=1}^{K} \gamma_{ij} = 0,$

(9) $\sum_{j=1}^{K} \gamma_{ij} = 0,$

(10) $\gamma_{ij} = \gamma_{ji} \qquad \forall i \neq j.$

The adding-up constraint (8) assures that the budget shares of all commodities sum to one. Homogeneity of degree zero (9) says that if all prices and income are multiplied by a positive constant k, the quantity demanded must remain unchanged. The symmetry constraint (10) deals with the substitution effect between commodities. The matrix of substitution effects is symmetric, meaning that the coefficient of the price of good i (ln p_i) has the same value in the budget share equation of good j (w_j) as the coefficient of ln p_j in w_i (Phlips 1974, Deaton and Muellbauer 1980b).

Banks et al. (1997) provide formulas for calculating price and budget elasticities of the commodities:

(11) $\quad \mu_i \equiv \frac{\partial w_i}{\partial \ln m} = \beta_i + \frac{2\lambda_i}{b(p)}\left\{\ln\left[\frac{m}{P(p)}\right]\right\}$

(12) $\quad \mu_{ij} \equiv \frac{\partial w_i}{\partial \ln p_j} = \gamma_{ij} - \mu_i\left(\alpha_j + \Sigma_k\ \gamma_{jk}\ln p_k\right) - \frac{\lambda_i\beta_j}{b(p)}\left\{\ln\left[\frac{m}{P(p)}\right]\right\}^2$

The budget elasticities e_i are given by

(13) $\quad e_i = \mu_i/w_i + 1$

The uncompensated (Marshallian) price elasticity e^u takes the income effect and the substitution effect into account and is computed by

(14) $\quad e_{ij}^u = \mu_{ij}/w_i - \delta_{ij}, \qquad where:\ \delta_{ij} = \begin{cases} 1 & if\ i = j \\ 0 & if\ i \neq j \end{cases}$.

The compensated (Hicksian) price elasticity e^c, which does not reflect the income effect, is computed by the following equation (13):

(15) $\quad e_{ij}^c = e_{ij}^u + e_i w_j$.

Household characteristics or socioeconomic variables (hc_h) may also be incorporated into the demand system and are included in the budget share equation (3) as follows:

(16) $\quad w_i = \alpha_i + \Sigma_{j=1}^{K}\gamma_{ij}\ln P_j + \beta_i \ln\left\{\frac{m}{P(p)}\right\} + \frac{\lambda_i}{b(p)}\left[\ln\left\{\frac{m}{P(p)}\right\}\right]^2 + \Sigma_{h=1}^{H}\theta_{ih}hc_h$.

Constraint (17) completes the above mentioned additivity restriction (8):

(17) $\quad \Sigma_{i=1}^{K}\theta_{ih} = 0$.

4.4 Unit values and zero observations

The data set, which was used for this study, contains the monthly food expenditures and quantities of nearly 12,000 households in Germany in the year 2003. As the prices of the commodities are not measured directly, these have to be derived from the average prices or unit values UV by dividing the expenditures E of a commodity by its quantity Q ($UV=E/Q$). By computing the unit values, differences in prices and differences in the quality are confounded. Cox and Wohlgenant (1986) developed a method to estimate quality adjusted prices, which is based on Houthakker (1952) and Theil (1952). The following hedonic price function is used:

$$(18) \quad UV_i = \delta_i + \textstyle\sum_c \kappa_{ic} C_{ic} + e_i,$$

where C_{ic} reflects several quality characteristics of commodity i ($c=1,...,n$). Quality adjusted prices p_i^* are then calculated by adding the constant δ_i and the error term e_i, which reflects the non-quality related price variations:

$$(19) \quad p_i^* = \delta_i + e_i.$$

As the quality characteristics are not given in the data set, socioeconomic household data may be used as proxy variables for the household preferences (Cox and Wohlgenant 1986). Quality adjusted prices based on household characteristics were for example estimated by Park et al. (1996), Park and Capps (1997), Dong et al. (1998), and Thiele (2001, 2008). In this study the following household characteristics are used to estimate the quality adjusted prices: number of persons living in the household, number of wage earners in the household, education level of the main income recipient, number of children under 18 years, and income. As the EVS 2003 survey was conducted throughout the whole year in Germany, price fluctuations due to seasonal or regional conditions are incorporated by dummy variables in the regression (Park et al. 1996, Thiele 2008).

A further specific characteristic of household data is the occurrence of zero observations of the dependent variables. Zero observations may arise because a consumer may not like a certain product, income restrictions may constrain the household to abstain from a product (corner solution), or the survey period may have been too short, so that a consumer did not buy a certain commodity in the time of the survey (Thiele 2008). If the zero observations are excluded from the estimation, a selectivity bias may appear (Maddala 1983). To adjust for the selectivity bias, a consistent-two-step-estimation (CTS) based on

Shonkwiler and Yen (1999) is conducted. The CTS-estimation was for example applied by Yen et al. (2002, 2003), Lambert et al. (2006), Akbay et al. (2007), and Thiele (2008).

In the first step, the likelihood that a household consumes commodity i is computed with a probit-analysis. Variables affecting the buying decision (z_i) are used to estimate the probability that a household consumes commodity i. The vector of variables z_i includes variables like income, commodity prices, number of persons and employees living in the household, age and education of the main income recipient and regional and seasonal variables. The results of the first step, the probability density function of the standard normal distribution $\phi(z_i\rho_i)$ and the cumulated density function of the standard normal distribution $\Phi(z_i\rho_i)$ are then included in the budget share equation of the QUAIDS and formula (16) transforms to:

(20) $\quad w_i^* = \Phi(z_i\rho_i)\{w_i\} + \sigma\,\phi(z_i\rho_i),$

where ρ_i is the vector of coefficients estimated by the probit-analyis and σ is a parameter to be estimated, which indicates if the estimation of the demand system without the correction factor for treating zero observations would have been biased (Thiele 2008).

The estimation of a demand system like (20) may lead to heteroskedasticity and a misspecified variance-covariance-matrix (Shonkwiler and Yen 1999), which is in this study corrected by the White-estimator (White 1980, Greene 2003).

Applying the CTS-estimation also leads to changed calculation formulas of the budget elasticities e_i and price elasticities e^u and e^c:

(21) $\quad e_i = \Phi(z_i\rho_i) * \mu_i/w_i + 1,$

(22) $\quad e_{ij}^u = \Phi(z_i\rho_i) * \mu_{ij}/w_i - \delta_{ij},\qquad where: \begin{pmatrix}\delta_{ij}=1 & if\ i=j \\ \delta_{ij}=0 & if\ i\neq j\end{pmatrix},$

(23) $\quad e_{ij}^c = e_{ij}^u + e_i w_j.$

4.5 Estimation strategy

The estimated food expenditures for each household that are obtained from (1) in the first stage are used as a regressor in the estimation of the quadratic almost ideal demand system (QUAIDS) instead of the observed expenditures. The purpose is to avoid the problem that the budget share of commodity i is directly calculated from the real expenditures. The estimation of the QUAIDS with the real observed expenditures can be inconsistent and biased, because of a likely correlation between the expenditure m and the error term in the budget share equation (Edgerton 1993, Henneberry et al. 1999, Zheng and Henneberry 2010).

The QUAIDS in this study is estimated in Stata with the nonlinear seemingly unrelated regression procedure of Poi (2008). During the estimation one of the six budget share equation is dropped (other food in stage 2 and breaded fish in stage 3) to avoid an error covariance matrix which is identically singular as the budget shares sum to one. Afterwards the missing parameters are calculated with the help of the additivity (8), homogeneity (9), and symmetry restrictions (10). Alternatively, the demand system could be estimated with all budget share equations to check if the restrictions (8) to (10) are fulfilled. As the household size may have an impact on the budget share (Abdulai 2002), this variable is included in the QUAIDS in stage 2 and 3. Figure 1 gives an overview on the estimation procedure and the commodity groups analyzed.

Figure 1: Schematic representation of the LES-QUAIDS-QUAIDS-System for fish consumption of German households

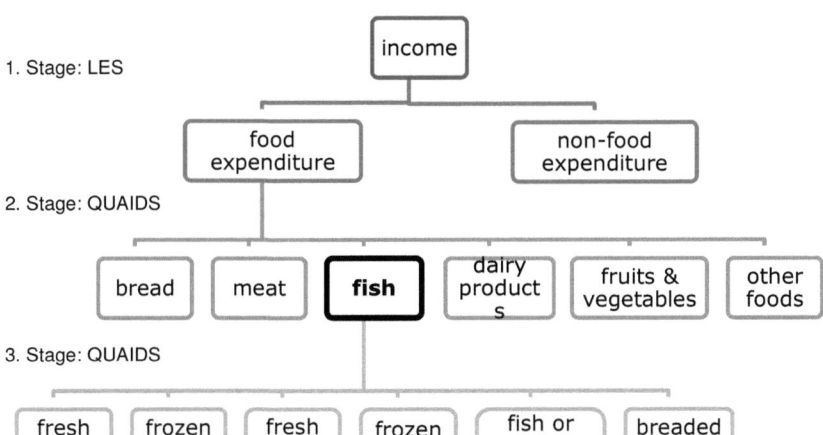

5. Results

In this section the expenditure and price elasticities of stage 2 and 3 will be presented. The estimation results of the LES in stage 1 and the QUAIDS in stage 2 and 3 can be found in the appendix. The estimated parameters are mostly statistically significant and also the coefficient σ, which indicates if the estimation of the demand system without the correction factor for treating zero observations would have been biased, is significant in 11 out of 12 demand equations. This might be a hint that the estimation would have been biased without this correction factor for the zero observations.

5.1 Results of stage 2: QUAIDS

All elasticities were calculated for the mean household. The expenditure elasticity for fish is about 1 while the lowest expenditure elasticity was found for bread with a value of ca. 0.75.

All own-price elasticities are statistically significant and negative. The uncompensated own-price elasticity for fish is -0.88 (see Table 3) and is thus only marginally higher than the elasticity for bread (-0.87) (see Table A 5). As the compensated elasticities do not reflect the income-effect, the compensated own-price elasticities are expected to be lower than the uncompensated elasticities: The own-price elasticity of fish decreases slightly to -0.84.

The uncompensated cross-price elasticities show substitutive and complementary relationships between the food commodity groups. Nearly all uncompensated cross-price elasticities are small or zero. The compensated cross-price elasticities are all positive, which means that all commodity groups can be seen as substitutes. In addition, the values are higher than is the case for the uncompensated cross-price elasticities. Dairy products and meat are the main substitutes for fish with compensated cross-price elasticities of 0.24 and 0.2, respectively (see Table A 5). The elasticities for the food commodity groups are given in Table A 5 in the appendix.

Table 3: Expenditure and price elasticities of fish

	Bread	Meat	Fish	Dairy products	Fruits and Vege-tables	other foods
expenditure elasticities	0.746 ***	1.120 ***	1.027 ***	0.926 ***	1.170 ***	0.982 ***
uncompen-sated price elasticity of fish	-0.001	-0.014 ***	-0.877 ***	0.019 ***	-0.020 ***	0.000
compensated price elasticity of fish	0.024 ***	0.023 ***	-0.843 ***	0.050 ***	0.019 ***	0.033 ***

*** indicates significance at p<0.01; ** indicates significance at p<0.05; * indicates significance at p<0.10

5.2 Results of stage 3: QUAIDS

The expenditure elasticities for fresh fish, frozen fish, and frozen fish fillets are nearly one, while the expenditure elasticities are lower for fresh fish fillets (0.8), fish or fish fillets without separation fresh or frozen (0.67), and breaded fish (0.19).

The uncompensated own-price elasticities are significantly above 1, meaning that a 1 percent price increase leads to a drop in demand of more than 1 percent. The highest uncompensated own-price elasticity was detected for breaded fish (-4.7) followed by frozen fish fillets (-3.35). The compensated own-price elasticities do not diverge much from the uncompensated ones.

Cross-price elasticities indicate that most fish products are seen as substitutive goods, while this is not the case for breaded fish, which is seen as complement to all other fish groups. The results of the cross-price elasticities for fish products are contrary to the results of meat demand of Thiele (2008), where nearly all meat products are seen as complementary goods. If the price of one fish product increases, the consumer seems to switch to other fish products and does not reduce its fish consumption in general. The elasticities of the QUAIDS for fish commodities are shown in Table 4.

Table 4: Expenditure elasticities and price elasticities of fish commodities

	fresh fish	frozen fish	fresh fish fillets	frozen fish fillets	fish or fish fillets without separation fresh or frozen	breaded fish
expenditure elasticities	0.997 ***	0.999 ***	0.795 ***	0.959 ***	0.665 ***	0.193 **
uncompensated price elasticities						
fresh fish	-1.860 ***	1.413 ***	1.122 ***	1.452 ***	1.135 ***	-0.886 ***
frozen fish	0.473 *	-3.345 ***	0.990 ***	0.975 ***	1.105 ***	1.363 ***
fresh fish fillets	0.792 ***	1.257 ***	-2.138 ***	1.096 ***	0.918 ***	-0.439 ***
frozen fish fillets	0.901 ***	1.690 ***	1.101 ***	-2.020 ***	1.098 ***	-0.362 *
fish or fish fillets without separation fresh or frozen	0.981 ***	0.834 ***	1.054 ***	1.083 ***	-1.764 ***	-0.416 **
breaded fish	-2.288 ***	-2.849 **	-3.159 ***	-3.591 ***	-3.540 ***	-4.702 ***
compensated price elasticities						
fresh fish	-1.772 ***	1.501 ***	1.192 ***	1.536 ***	1.194 ***	-0.869 ***
frozen fish	0.500 *	-3.318 ***	1.012 ***	1.001 ***	1.123 ***	1.368 ***
fresh fish fillets	0.879 ***	1.345 ***	-2.068 ***	1.180 ***	0.976 ***	-0.422 ***
frozen fish fillets	0.956 ***	1.746 ***	1.145 ***	-1.967 ***	1.135 ***	-0.352 *
fish or fish fillets without separation fresh or frozen	1.130 ***	0.983 ***	1.173 ***	1.226 ***	-1.664 ***	-0.387 **
breaded fish	-2.194 ***	-2.755 **	-3.084 ***	-3.501 ***	-3.478 ***	-4.684 ***

*** indicates significance at p<0.01; ** indicates significance at p<0.05; * indicates significance at p<0.10

5.3 Comparison of results

The results of stage 2 can be compared to other (fish) demand studies only with restraints, as the aggregated food commodity groups are not identical and also the method chosen is not the same. Only a few studies calculated own-price elasticities of fish in Germany explicitly. In this study the uncompensated own-price elasticity of fish is estimated to be around -0.88 and the compensated own-price elasticity is -0.84. These values are in line with the results of Henning and Michalek (1992), Thiele (2001) and Wildner (2001a+b), and Sommer (1985). The results of Ryll (1984), Michalek and Keyzer (1992) and Wildner and von Cramon-Taubadel (2000) show larger discrepancies compared to the estimated demand elasticity of this study.

The expenditure elasticity of fish was estimated to be around 1 in this study. This value is above the income or expenditure elasticities calculated in earlier studies by Hoffmann (2003), Thiele (2001), Wildner (2001a+b), Henning and Michalek (1992), and Michalek and Keyzer (1992) (see Table 1).

The results of the third stage are not comparable to other studies because the demand for fish in Germany was not analyzed in detail. Only Ryll (1984) and Sommer (1985) made detailed analysis, but the fish commodity groups are not comparable between these studies.

Compared to international meta-analysis of fish demand, it can be stated that the results of this study are in line with several other studies. The demand elasticity for fish in Germany was estimated to be about -0.88. Gallet (2009) reviewed 168 studies published between 1951 and 2007 and finds that the median elasticity of fish is -0.8. Andreyeva et al. (2010) reviewed 18 US-based studies and compute a mean price elasticity of -0.5 with a 95-percent confidence interval ranging from -0.3 to -0.7.

Asche et al. (2005) find that substitutes of seafood products tend to be other seafood product and that the degree of substitution between fish and meat is substantially less. This is in line with the results from stage 2 and 3. Asche et al. (2005) also find that demand in most markets seems to be elastic. As a large part of the review focuses on individual fish species, this result may be comparable to the results of stage 3, which also showed elastic demand for the fish commodities.

5.4   Sensitivity of the results

The quality of the estimation results may crucially depend on modeling choices, assumptions, types of data and statistical techniques (Okrent and Alston 2011, Gallet 2009). The sensitivity of the QUAIDS estimation results is tested regarding changes in the estimation strategy, namely computing the coefficients and elasticities with and without the consistent two step estimation procedure (CTS) and quality adjusted prices (QAP) (see Section 4.4).

The sensitivity analysis in this section is constrained to the Marshallian price elasticities for fish. Detailed results of the estimated coefficients and elasticities can be found in Table A 3 to Table A 20 in the appendix.

The own-price elasticity of fish is -0.88 when the QUAIDS is estimated with the CTS and quality adjusted prices. Estimating the QUAIDS without these two modifications leads to a drop in the elasticity to -0.5. Estimating the demand system with the CTS and without quality adjusted prices, the own-price elasticity of fish is -0.25. Estimating the system with quality adjusted prices but without the CTS, then the elasticity is calculated to be -0.96 (see Table 5). In contrast to fish, the elasticities for all other food commodity in stage two do not vary much. Compared to the other food groups, the share of zero consumption of fish is relatively high (ca. 23.7 percent, see Table A 1). This might be a reason, why the elasticity of fish reacts sensitively to changes in the estimation method. However, the qualitative results do not change: in all four cases the own-price elasticity is inelastic.

In the case of the fish commodities the estimation method has a huge impact on the calculated elasticities. In the estimation with CTS and quality adjusted prices the own-price elasticity of fresh fish is -1.86. If the system is estimated without CTS and quality adjusted prices the elasticity is -1.59. In the estimation with quality adjusted prices but without CTS the elasticity is calculated to be -2.29. In the case of the estimation with CTS but without quality adjusted prices leads to questionable positive elasticity of 1.09. This is also the case for four other fish products. Table 5 summarizes the impact of the estimation method on the Marshallian own-price elasticity.

Table 5: Sensitivity of the Marshallian own-price elasticity to changes in the estimation method

	Estimation of the QUAIDS...			
	with CTS and QAP	with CTS	with QAP	without CTS and QAP
Fish (stage 2)	-0.877	-0.245	-0.963	-0.495
fresh fish	-1.860	1.087	-2.291	-1.588
frozen fish	-3.345	3.943	-1.926	-1.198
fresh fish fillets	-2.138	1.269	-8.473	-1.052
frozen fish fillets	-2.020	3.725	-2.408	-0.331
fish or fish fillets without separation fresh or frozen	-1.764	0.107	-2.171	-0.832
breaded fish	-4.702	-4.449	-1.392	-0.882

CTS: Consistent-two-step estimation (Shonkwiler and Yen 1999)
QAP: Quality-adjusted prices (Cox and Wohlgenant 1986)

6. Discussion and Summary

The demand for fish in Germany is relatively low compared to the average of EU-27-countries. The share of the expenditures for fish on the total food expenditures is also relatively small with 3.4 percent. Since 1985 the demand for fish in Germany was not analyzed in detail. In a multistage framework a QUAIDS was estimated. Expenditure elasticities and price elasticities were computed for food groups and for six fish commodity groups. Cross sectional household data of the year 2003 were used. Quality adjusted prices and a consistent two step estimation procedure were used to estimate the QUAIDS. The significance of the sigma–coefficient in the demand estimations indicates that the estimation would have been biased without this correction term. However, the sensitivity of the estimation and elasticity results to modifications in the estimation procedure was tested.

Compared to other food products the share of consumers is significantly lower for fish products (see Table A 1 in the appendix). The problem of zero observations increases dramatically when the six fish sub-groups are regarded, where the share of consumers is only between 4 percent for frozen fish and 18.4 percent for fish or fish fillets without separation fresh or frozen. The relatively low number of fish consumers might limit the validity of the results of stage 3. This might also be a reason for the violation of the homogeneity restriction in stage 3, as the sum of the compensated price elasticities by each column is not zero.

A QUAIDS was chosen for the estimation of food and fish demand because Schmitz (2007) showed that Engel-curves are nonlinear for food demand in Germany. In stage 2 three out of six λ-coefficients are statistically significant while in stage 3 this is the case for each fish product. Nevertheless, the values of the λ-coefficients indicate that the quadratic expenditure term has only a marginal effect on the budget share for each analyzed commodity.

The estimated price coefficients of the QUAIDS are significant and thus indicate that price policy can be an important policy instrument in the case of food and fish commodities. The results further indicate that the household size has a small but statistically significant influence on the consumption pattern, which is slightly negative for fish and fruits and vegetables and positive for the other four food groups. In the case of the six fish commodities the influence is negative with the exception in the demand for breaded fish.

The results on the elasticities show that the own-price elasticity of fish is inelastic and its value lies around -0.8. The expenditure elasticity is slightly above one. The expenditure elasticities of the six fish sub-groups lie near or below one. The price elasticities are all elastic, where the elasticity for breaded fish is highest while the lowest elasticities were found for fresh fish and fish without separation fresh or frozen.

A re-estimation of the QUAIDS with and without the consistent two step estimation and quality adjusted prices showed that especially the fish price elasticities react much more sensitively to changes in the estimation method than is the case for other food commodities. Nevertheless, the qualitative results stay the same: fish demand is inelastic in Germany. The selection of the estimation method also has huge impacts on the own-price elasticities of the fish products of the third estimation stage.

As the QUAIDS is based on the neoclassical assumption that the consumer maximizes its utility, the results have to be interpreted with care. Only the choices of the consumers are observable (Okrent and Alston 2011), but the unobservable part is, if utility is really maximized.

This study only sheds some light on several fish product forms in the year 2003. Further research would be useful to gain information on how consumers react to price changes for a certain fish species and if there are differences in consumer behavior regarding the origin of a fish (aquaculture or capture fisheries). Further research on the cross-price elasticities is also useful to get more and detailed information on the relationships between the several fish products as well as the food commodity groups. Trends or changes in the reaction to price changes can hardly be measured within the data set of one year. Time-series analysis could be useful to get information on the development and changes of consumers' behavior. Additionally, demand analysis by different household types could gain more information on the structure of fish demand. The area of real field experiments is also an interesting way to learn more on consumers' choices in respect to fish.

References

Abdulai, A. (2002): Household demand for food in Switzerland. A quadratic almost ideal demand system. Swiss Journal of Economics and Statistics 138(1): 1-18.

Akbay, C., Boz, I. and Chern, W.S. (2007): Household food consumption in Turkey. European Review of Agricultural Economics 34(2): 209-231.

Andreyeva, T., Long, M.W. and Brownell, K.D. (2010): The impact of food prices on consumption: a systematic review of research on the price elasticity of demand for food. American Journal of Public Health 100(2): 216-222.

Asche, F., Bjørndal, T. and Gordon, D.V. (2005): Demand structure for fish. SNF Working Paper No. 37/05, Institute for Research in Economics and Business Administration, Bergen, August 2005.

Barnett, W.A. and Serletis, A. (2008): Consumer preferences and demand systems. Journal of Econometrics 147(2): 210-224.

Banks, J., Blundell, R. and Lewbel, A. (1997): Quadratic Engel curves and consumer demand. The Review of Economics and Statistics 79(4): 527-539.

Bjørndal, T., Salvanes, K.G. and Andreassen, J.H. (1992): The demand for salmon in France: the effects of marketing and structural change. Applied Economics 24(9): 1027-1034.

Blundell, R., Pashardes, P. and Weber, G. (1993): What do we learn about consumer demand patterns from micro data? American Economic Review 83(3): 570-597.

Chavas, J.P. (1983): Structural change in the demand for meat. American Journal of Agricultural Economics 65(1): 148-153.

Cox, T.L. and Wohlgenant, M.K. (1986): Prices and quality effects in cross-sectional demand analysis. American Journal of Agricultural Economics 68(4): 908-919.

Czajka, S. and Kott, K. (2006): Konsumausgaben privater Haushalte für Nahrungsmittel, Getränke und Tabakwaren 2003. Wirtschaft und Statistik 6/2006: 630-643.

Deaton, A. (1986): Demand analysis. In: Griliches, Z. and Intrilligator, M.D. (eds.): Handbook of econometrics 3: 1767-1839. Elsevier Science Publishers B.V., Amsterdam.

Deaton, A. and Muellbauer, J. (1980a): An almost ideal demand system. The American Economic Review 70(3): 312-326.

Deaton, A. and Muellbauer, J. (1980b): Economics and consumer behavior. Cambridge University Press, Cambridge.

Destatis (2005): Einkommens- und Verbrauchsstichprobe – Aufgabe, Methode und Durchführung der EVS. Fachserie 15 Wirtschaftsrechnungen, Heft 7. Statistisches Bundesamt, Wiesbaden.

Destatis (2011): Statistisches Jahrbuch 2011. Statistisches Bundesamt, Wiesbaden.

Dey, M.M. (2000): Analysis of demand for fish in Bangladesh. Aquaculture Economics & Management 4(1-2): 65-83.

Dey, M.M., Briones, R.M. and Ahmed, M. (2005): Disaggregated analysis of fish supply, demand, and trade in Asia: baseline model and estimation strategy. Aquaculture Economics & Management 9(1-2): 113-139.

Dey, M.M., Garcia, Y.T., Kumar, P., Piumsombun, S., Haque, M.S., Li, L., Radam, A., Senaratne, A., Khiem, N.T. and Koeshendrajana, S. (2008): Demand for fish in Asia: a cross- country analysis. The Australian Journal of Agricultural and Resource Economics 52(3), 321-338.

Dey, M.M., Alam, M.F. and Paraguas, F.J. (2011): A multistage budgeting approach to the analysis of demand for fish: an application to inland areas of Bangladesh. Marine Resource Economics 26(1): 35-58.

Dong, D., Shonkwiler, J.S. and Capps, O. (1998): Estimation of demand functions using cross-sectional household data: the problem revisited. American Journal of Agricultural Economics 80(3): 466-473.

Ecker, O. and Qaim, M. (2011): Analyzing nutritional impacts of policies: an empirical study for Malawi. World Development 39(3): 412-428.

Edgerton, D.L. (1993): On the estimation of separable demand models. Journal of Agricultural and Resource Economics 18(2): 141-146.

Edgerton, D.L. (1997): Weak separability and the estimation of elasticities in multistage demand systems. American Journal of Agricultural Economics 79(1): 62-79.

EU (2012): Facts and figures on the common fisheries policy – Basic statistical data – 2012 edition. Publications Office of the European Union, Luxembourg.

FAO (2012): Food Outlook, May 2012. Food and Agriculture Organization of the United Nations, Rome.

FAOSTAT (2012): FAOSTAT <http://faostat3.fao.org/home/index.html>.

FAO/WHO (2011): Report of the joint FAO/WHO expert consultation on the risks and benefits of fish consumption. Food and Agriculture Organization of the United Nations, Rome, World Health Organization, Geneva.

Gallet, C.A. (2009): The demand for fish: a meta-analysis of the own-price elasticity. Aquaculture Economics & Management 13(3): 235-245.

Gao, X.M. Wailes, E.J. and Cramer, G.L. (1996): A two-stage rural household demand analysis: microdata evidence from Jiangsu Province, China. American Journal of Agricultural Economics 78(3): 604-613.

Greene, W. (2003): Econometric Analysis. 5th edition, Prentice Hall, New Jersey.

Henneberry, S.R., Piewthongngam, K. and Qiang, H. (1999): Consumer food safety concerns and fresh produce consumption. Journal of Agricultural and Resource Economics 24(1): 98-113.

Henning, C. and Michalek, J. (1992): Innovatives Konsumverhalten für Nahrungsmittel? Ableitung und Schätzung eines auf Nahrungsmittel fokussierten kompletten Nachfragesystems unter Berücksichtigung von zeitlichen Präferenzänderungen. Agrarwirtschaft 41(11): 330-342.

Hoffmann, C. (2003): Die Nachfrage nach Nahrungs- und Genußmitteln privater Haushalte vor dem Hintergrund zukünftiger Rahmenbedingungen. Studien zur Haushaltsökonomie 28. Peter Lang, Frankfurt on the Main.

Houthakker, H.S. (1952): Compensated changes in quantities and qualities consumed. The Review of Economic Studies 19(3): 155-164.

Intriligator, M.D. (1978): Econometric models, techniques, and applications. North-Holland Publishing Company, Amsterdam, Oxford.

Kumar, P., Dey, M.M. and Paraguas, F.J. (2005): Demand for fish by species in India: three-stage budgeting framework. Agricultural Economics Research Review 18(2): 167-186.

Lambert, R., Larue, B., Yélou, C. and Criner, G. (2006): Fish and meat demand in Canada: regional differences and weak separability. Agribusiness 22(2): 175-199.

Lau, L.J. (1986): Functional forms in econometric model building. In: Griliches, Z. and Intrilligator, M.D. (eds.): Handbook of econometrics 3: 1515-1566. Elsevier Science Publishers B.V., Amsterdam.

Lewbel, A. (1991): The rank of demand systems: Theory and nonparametric estimation. Econometrica 59(3): 711-730.

Maddala, G.S. (1983): Limited-dependent and qualitative variables in econometrics. Cambridge University Press, Cambridge.

Michalek, J. and Keyzer, M.A. (1992): Estimation of a two-stage LES-AIDS consumer demand system for eight EC countries. European Review of Agricultural Economics 19(2): 137-163.

Mozzafarin, D. and Rimm, E.B. (2006): Fish intake, contaminants, and human health: evaluating the risks and the benefits. Journal of the American Medical Association 296(15): 1885-1899.

Okrent, A.M. and Alston, J.M. (2011): Demand for food in the United States: A review of literature, evaluation of previous estimates, and presentation of new estimates. Giannini Foundation Monograph 48.

Pan, S., Mohanty, S. and Welch, M. (2008): India edible oil consumption: a censored incomplete demand approach. Journal of Agricultural and Applied Economics 40(3): 821-835.

Park, J.L. and Capps, O. (1997): Demand for prepared meals by U.S. households. American Journal of Agricultural Economics 79(3): 814-824.

Park, J.L., Holcomb, R.B., Raper, K.C. and Capps, O. (1996): A demand system analysis of food commodities by U.S. households segmented by income. American Journal of Agricultural Economics 78(2): 290-300.

Pawlik, H. (1993): Die Nachfrage nach Tiefkühlkost – Struktur, Bestimmungsgründe und Perspektiven. Agrarmarkt-Studien, Heft 39. Verlag Paul Parey, Hamburg, Berlin.

Perk, J., De Backer, G., Gohlke, H., Graham, I., Reiner, Z., Verschuren, M., Albus, C., Benlian, P., Boysen, G., Cifkova, R., Deaton, C., Ebrahim, S., Fisher, M., Germano, G., Hobbs, R., Hoes, A., Karadeniz, S., Mezzani, A., Prescott, E., Ryden, L., Scherer, M., Syvänne, M., Scholte Op Reimer, W.J.M., Vrints, C., Wood, D., Zamorano, J.L. and Zannad, F. (2012): European guidelines on cardiovascular disease prevention in clinical practice (version 2012). European Heart Journal 33: 1635-1701.

Phlips, L. (1974): Applied consumption analysis. North Holland Publishing Company, Amsterdam, Oxford.

Poi, B.P. (2008): Demand system estimation: update. The Stata Journal 8(4): 554-556.

Ryll, E. (1984): Bestimmungsgründe und Elastizitäten der Nachfrage nach Fisch und Fischwaren in der Bundesrepublik Deutschland. Berichte über Landwirtschaft 62: 208-221. Paul Parey, Hamburg, Berlin.

Salvanes, K.G. and De Voretz, D.J. (1997): Household demand for fish and meat products: separability and demographic effects. Marine Resource Economics 12(1): 37-55.

Schmitz, S.M. (2007): Nachfrageanalyse, Economies of Scale und Preisdifferenzierung im deutschen Lebensmitteleinzelhandel. Dissertation, University of Kiel.

Schons, H.-P. (1993): Vorschätzung des Nahrungsmittelverbrauchs in den Mitgliedsländern der EG (12) und ausgewählten Drittländern für die Zieljahre 1995 und 2000. Schriftenreihe des Bundesministers für Ernährung, Landwirtschaft und Forsten, Reihe A: Angewandte Wissenschaft, Heft 421. Landwirtschaftsverlag GmbH, Münster.

Shonkwiler, J.S. and Yen, S.T. (1999): Two-step estimation of a censored system of equations. American Journal of Agricultural Economics 81(4): 972-982.

Sommer, U. (1985): Quantitative Analyse der Nachfrage nach Fisch und Fischwaren in der Bundesrepublik Deutschland 1965 bis 1981. Berichte über Landwirtschaft 63: 82-101. Paul Parey, Hamburg, Berlin.

Theil, H. (1952): Qualities, prices and budget enquiries. The Review of Economic Studies 19(3): 129-147.

Thiele, S. (2001): Ausgaben- und Preiselastizitäten der Nahrungsmittelnachfrage auf Basis von Querschnittsdaten: Eine Systemschätzung für die Bundesrepublik Deutschland. Agrarwirtschaft 50(2): 108-115.

Thiele, S. (2008): Elastizitäten der Nachfrage privater Haushalte nach Nahrungsmitteln – Schätzung eines AIDS auf Basis der Einkommens- und Verbrauchsstichprobe 2003. Agrarwirtschaft 57(5): 258-268.

White, H. (1980): A heteroscedasticity consistent covariance matrix and a direct test for heteroscedasticity. Econometrica 46(4): 817-838.

Wildner, S. (2001a): Quantifizierung der Preis- und Ausgabenelastizität für Nahrungsmittel in Deutschland: Schätzung eines LA/AIDS. Agrarwirtschaft 50(5): 275-285.

Wildner, S. (2001b): Die Nachfrage nach Nahrungsmitteln in Deutschland unter besonderer Berücksichtigung von Gesundheitsinformationen. Agrarwirtschaft, Sonderheft 169, AgriMedia, Bergen/Dumme.

Wildner, S. and von Cramon-Taubadel, S. (2000): Die Bedeutung von Veränderungen der Nachfrage für die Wettbewerbsfähigkeit des Agrarsektors: Erste Ergebnisse einer neuen Nachfrageschätzung. In: von Alvensleben, R., Koester, U. and Langbehn, C. (eds.): Schriften der Gesellschaft für Wirtschafts- und Sozialwissenschaften des Landbaus e.V., 36: 63-74, Landwirtschaftsverlag GmbH, Münster-Hiltrup.

Yen, S.T., Kan, K. and Su, S.J. (2002): Household demand for fats and oils: two step estimation of a censored demand system. Applied Economics 34(14): 1799-1806.

Yen, S.T., Lin, B.H. and Smallwood, D.M. (2003): Quasi- and simulated-likelihood approaches to censored demand systems: food consumption by food stamp recipients in the United States. American Journal of Agricultural Economics 85(2): 458-478.

Zheng, Z. and Henneberry, S.R. (2010): An analysis of food grain consumption in urban Jiangsu Province, China. Journal of Agricultural and Applied Economics 42(2): 337-355.

Appendix

Table A 1: Definition and descriptive statistics of used variables

		mean	std.-dev.
	quality adjusted price (€/kg)	2.23	0.858
Bread	budget share	0.1905	0.084
	share of consumers	0.9988	
	quality adjusted price (€/kg)	6.42	2.655
Meat	budget share	0.2343	0.117
	share of consumers	0.9814	
	quality adjusted price (€/kg)	6.05	3.286
Fish	budget share	0.0334	0.041
	share of consumers	0.7633	
	quality adjusted price (€/kg)	3.46	1.655
Dairy products	budget share	0.1675	0.070
	share of consumers	0.9978	
	quality adjusted price (€/kg)	1.8	0.737
Fruits and Vegetables	budget share	0.2153	0.099
	share of consumers	0.9965	
	quality adjusted price (€/kg)	4.09	2.010
other foods	budget share	0.1590	0.076
	share of consumers	0.9965	
	quality adjusted price (€/kg)	11.29	2.309
fresh fish	budget share	0.1760	0.351
	share of consumers	0.1171	
	quality adjusted price (€/kg)	7.09	0.748
frozen fish	budget share	0.0547	0.206
	share of consumers	0.0401	
	quality adjusted price (€/kg)	8.53	1.614
fresh fish fillets	budget share	0.1740	0.348
	share of consumers	0.1175	
	quality adjusted price (€/kg)	5.93	0.907
frozen fish fillets	budget share	0.1103	0.285
	share of consumers	0.0772	
fish or fish fillets without	quality adjusted price (€/kg)	7.82	2.098
separation fresh or	budget share	0.2975	0.428
frozen	share of consumers	0.1838	
	quality adjusted price (€/kg)	4.78	0.881
breaded fish	budget share	0.1876	0.353
	share of consumers	0.1434	

Table A 1 continued: Definition and descriptive statistics of used variables

		mean	std.-dev.
income	net household income per month (€)	3498.97	2118.724
age	age of main income recipient	51.8	13.773
number of persons in the household		2.4210	1.212
number of children under 18 years		0.4839	0.877
number of wage earners in the household		1.0647	0.886
single, men	Dummy =1 if main income recipient is single male, else=0	0.1480	
low education	Dummy =1 if main income recipient has no training qualification and is not in job training, else=0	0.0204	
high education	Dummy =1 if main income recipient has an universtity degree or an university of applied science degree, else=0	0.3316	
north	Dummy =1 if household is in Schleswig-Holstein, Lower Saxony, Hamburg or Bremen, else =0	0.1609	
south	Dummy =1 if household is in Hesse, Rhineland-Palatinate, Baden-Württemberg, Bavaria or Saarland, else=0	0.4055	
east	Dummy =1 if household is in Brandenburg, Mecklenburg-West Pomerania, Saxony, Saxony-Anhalt, Thuringia, else=0	0.1983	
first quarter	Dummy =1 if household recorded expenditures in January, February or March, else =0	0.2325	
second quarter	Dummy =1 if household recorded expenditures in April, May or June, else =0	0.2564	
fourth quarter	Dummy =1 if household recorded expenditures in October, November or December, else =0	0.2607	

Table A 2: Constrained OLS regression results of the LES

	Coef.	Std. Err.	t	P>|t|	95% Conf. Interval	
Stone Price Index Food	-0.343	0.009	-40.20	0	-0.359	-0.326
ln expenditure non-food	-4.256	0.068	-62.64	0	-4.389	-4.122
ln income	4.598	0.072	63.47	0	4.456	4.740
household size	0.193	0.004	46.06	0	0.185	0.202
Dummy north	0.032	0.012	2.60	0.009	0.008	0.055
Dummy south	0.007	0.010	0.71	0.48	-0.012	0.026
Dummy east	-0.054	0.012	-4.70	0	-0.077	-0.032
Dummy high education	0.073	0.008	8.81	0	0.057	0.090
Dummy low education	-0.139	0.027	-5.17	0	-0.192	-0.086
single household male	-0.193	0.012	-15.43	0	-0.217	-0.168
age	0.007	0.000	24.12	0	0.006	0.008
constant	1.745	0.058	30.29	0	1.632	1.857

Table A 3: QUAIDS estimation results food commodity groups

	1	2	3	4	5	6
	Bread	Meat	Fish	Dairy products	Fruits and Vegetables	other foods
α_i	-0.8789 ***	0.3806 ***	0.0802 ***	-0.3631 ***	1.0000 ***	0.7812 ***
	(-5.97)	(9.57)	(11.46)	(-4.37)	(9.13)	(7.64)
β_i	-0.0449 ***	0.029 ***	0.0004	-0.0134 ***	0.0337 ***	-0.0048
	(-12.07)	(6.83)	(0.29)	(-5.05)	(9.05)	(-1.29)
γ_{i1}	0.0830 ***	-0.0149 ***	-0.0052 ***	0.0039	-0.0419 ***	-0.0248 ***
	(13.27)	(-5.32)	(-4.56)	(1.51)	(-9.49)	(-7.99)
γ_{i2}		0.0275 ***	-0.0011	0.0114 ***	-0.0047	-0.0182 ***
		(6.6)	(-0.89)	(4.94)	(-1.62)	(-7.47)
γ_{i3}			0.0054 ***	0.0025 ***	-0.0014	-0.0002
			(6.03)	(2.64)	(-1.11)	(-0.23)
γ_{i4}				0.0141 ***	-0.0166 ***	-0.0153 ***
				(5.91)	(-6.48)	(-8.07)
γ_{i5}					0.0702 ***	-0.0056 **
					(13.74)	(-2.05)
γ_{i6}						0.0144 **
						(2.51)
λ_i	-0.0021 ***	0.0008 **	0.0001	-0.0002	0.0016 ***	0.0002
	(-9.05)	(2.26)	(1.14)	(-1.23)	(6.34)	(0.75)
θ_i	0.0147 ***	0.0025 *	-0.0064 ***	0.0058 ***	-0.0263 ***	0.0096 ***
	(-15.12)	(1.87)	(-14.62)	(7.12)	(-22.96)	(10.41)
σ_i	4.1859 ***	-0.7013 ***	-0.0695 ***	1.9778 ***	-2.5349 ***	-2.8581 ***
	(8.31)	(-5.85)	(-5.49)	(6.82)	(-6.72)	(-7.68)

*** indicates significance at p<0.01; ** indicates significance at p<0.05; * indicates significance at p<0.10

z-statistics in parantheses

Table A 4: QUAIDS estimation results of fish commodities

	1	2	3	4	5	6
	fresh fish	frozen fish	fresh fish fillets	frozen fish fillets	fish or fish fillets without separation fresh or frozen	breaded fish
α_i	1.5364 ***	2.7153 ***	1.0081 ***	1.8041 ***	0.6229 ***	-6.6868 ***
	(-13.75)	(-7.41)	(7.93)	(8.05)	(4.55)	(-13.01)
β_i	0.0243 ***	0.0081	-0.0043	0.0136	-0.0359 **	-0.0059
	(2.73)	(1.04)	(-0.42)	(1.6)	(-2.29)	(-0.16)
γ_{i1}	-0.1384 ***	0.0754 ***	0.1271 ***	0.1452 ***	0.1562 ***	-0.3655 ***
	(-2.75)	(4.64)	(7.97)	(9.68)	(11.77)	(-8.46)
γ_{i2}		-0.1246 **	0.0667 ***	0.0899 ***	0.0436 ***	-0.151 ***
		(-2.03)	(5.3)	(4.34)	(4.48)	(-3.22)
γ_{i3}			-0.218 ***	0.1093 ***	0.1417 ***	-0.2268 ***
			(-4.82)	(8.68)	(13.24)	(-5.35)
γ_{i4}				-0.1145 *	0.1083 ***	-0.3381 ***
				(-1.77)	(9.39)	(-5.98)
γ_{i5}					-0.2679 ***	-0.1819 ***
					(-6.62)	(-3.88)
γ_{i6}						0.5323 ***
						(5.00)
λ_i	0.003 ***	0.001 **	0.0035 ***	0.0022 ***	0.0063 ***	0.016 ***
	(3.86)	(2.23)	(4.13)	(3.96)	(4.3)	(4.5)
θ_i	-0.034 ***	-0.0136 ***	-0.0263 ***	-0.0148 **	-0.0392 ***	0.1278 ***
	(-5.56)	(-3.03)	(-4.18)	(-2.46)	(-5.19)	(17.32)
σ_i	-1.4338 ***	-3.3778 ***	-0.9179 ***	-2.2671 ***	-0.2218	8.2186 ***
	(-8.33)	(-7.37)	(-5.3)	(-8.26)	(-1.16)	(13.15)

*** indicates significance at p<0.01; ** indicates significance at p<0.05; * indicates significance at p<0.10

z-statistics in parantheses

Table A 5: Expenditure elasticities and price elasticities of food commodity groups (estimation with quality adjusted prices and consistent two step estimation)

	Bread	Meat	Fish	Dairy products	Fruits and Vege-tables	other foods
expenditure elasticities	0.746 ***	1.120 ***	1.027 ***	0.926 ***	1.170 ***	0.982 ***
uncompensated price elasticities						
Bread	-0.867 ***	0.057 ***	-0.096 **	-0.048 ***	-0.007	-0.147 ***
Meat	0.039 ***	-0.951 ***	-0.037	0.087 ***	-0.089 ***	-0.089 ***
Fish	-0.001	-0.014 ***	-0.877 ***	0.019 ***	-0.020 ***	0.000
Dairy products	-0.072 ***	0.083 ***	0.068 **	-0.955 ***	-0.005	-0.087 ***
Fruits and Vegetables	0.063 ***	-0.134 ***	-0.058	-0.011	-0.893 ***	-0.012
other foods	0.074 ***	-0.152 ***	-0.025	-0.023 **	-0.145 ***	-0.911 ***
compensated price elasticities						
Bread	-0.725 ***	0.270 ***	0.100 *	0.129 ***	0.216 ***	0.040 ***
Meat	0.214 ***	-0.689 ***	0.203 ***	0.304 ***	0.185 ***	0.141 ***
Fish	0.024 ***	0.023 ***	-0.843 ***	0.050 ***	0.019 ***	0.033 ***
Dairy products	0.053 ***	0.271 ***	0.240 ***	-0.800 ***	0.191 ***	0.078 ***
Fruits and Vegetables	0.224 ***	0.107 **	0.163 ***	0.188 ***	-0.641 ***	0.199 ***
other foods	0.193 ***	0.026 **	0.139 ***	0.124 ***	0.041 ***	-0.755 ***

*** indicates significance at p<0.01; ** indicates significance at p<0.05; * indicates significance at p<0.10

Table A 6: Expenditure elasticities and price elasticities of food commodity groups (estimation without quality adjusted prices and consistent two step estimation)

	Bread	Meat	Fish	Dairy products	Fruits and Vege-tables	other foods
expenditure elasticities	0.684 ***	1.194 ***	0.869 ***	0.844 ***	1.286 ***	0.940 ***
uncompensated price elasticities						
Bread	-0.703 ***	-0.044 ***	-0.214 ***	-0.016 *	-0.127 ***	-0.065 ***
Meat	0.053 ***	-0.931 ***	-0.095 ***	0.147 ***	-0.140 ***	-0.119 ***
Fish	-0.028 ***	-0.026 ***	-0.495 ***	0.008	-0.038 ***	0.006
Dairy products	0.013	0.050 ***	0.029	-0.931 ***	-0.080 ***	-0.068 ***
Fruits and Vegetables	-0.006	-0.123 ***	-0.131 ***	-0.003	-0.821 ***	-0.032 ***
other foods	-0.024 ***	-0.113 ***	0.033	-0.055 ***	-0.070 ***	-0.831 ***
compensated price elasticities						
Bread	-0.573 ***	0.183 ***	-0.049 *	0.144 ***	0.118 ***	0.114 ***
Meat	0.213 ***	-0.652 ***	0.108 ***	0.345 ***	0.162 ***	0.102 ***
Fish	-0.005	0.014 ***	-0.466 ***	0.036 ***	0.005	0.038 ***
Dairy products	0.128 ***	0.250 ***	0.174 ***	-0.789 ***	0.135 ***	0.089 ***
Fruits and Vegetables	0.141 ***	0.134 ***	0.056 **	0.179 ***	-0.545 ***	0.170 ***
other foods	0.085 ***	0.077 ***	0.171 ***	0.079 ***	0.134 ***	-0.682 ***

*** indicates significance at p<0.01; ** indicates significance at p<0.05; * indicates significance at p<0.10

Table A 7: Expenditure elasticities and price elasticities of food commodity groups (estimation without quality adjusted prices but with consistent two step estimation)

	Bread	Meat	Fish	Dairy products	Fruits and Vegetables	other foods
expenditure elasticities	0.695 ***	1.181 ***	0.484 ***	0.867 ***	1.243 ***	1.017 ***
uncompensated price elasticities						
Bread	-0.806 ***	0.036 **	-0.336 ***	-0.041 ***	-0.034 **	-0.123 ***
Meat	0.029 **	-0.930 ***	-0.029	0.120 ***	-0.120 ***	-0.094 ***
Fish	0.023 ***	-0.052 ***	-0.245 ***	0.034 ***	-0.079 ***	-0.049 ***
Dairy products	-0.075 ***	0.113 ***	-0.086 **	-0.962 ***	-0.001	-0.100 ***
Fruits and Vegetables	0.039 **	-0.166 ***	0.072 *	-0.007	-0.851 ***	-0.005
other foods	0.059 ***	-0.160 ***	0.078 ***	-0.026 **	-0.129 ***	-1.116 ***
compensated price elasticities						
Bread	-0.674 ***	0.260 ***	-0.243 ***	0.124 ***	0.203 ***	0.071 ***
Meat	0.192 ***	-0.653 ***	0.084 **	0.323 ***	0.172 ***	0.144 ***
Fish	0.046 ***	-0.013 **	-0.229 ***	0.063 ***	-0.038 ***	-0.015 *
Dairy products	0.041 **	0.311 ***	-0.005	-0.817 ***	0.208 ***	0.071 ***
Fruits and Vegetables	0.188 ***	0.088 ***	0.177 ***	0.180 ***	-0.584 ***	0.214 ***
other foods	0.170 ***	0.028 **	0.155 ***	0.112 ***	0.068 ***	-0.954 ***

*** indicates significance at p<0.01; ** indicates significance at p<0.05; * indicates significance at p<0.10

Table A 8: Expenditure elasticities and price elasticities of food commodity groups (estimation with quality adjusted prices but without consistent two step estimation)

	Bread	Meat	Fish	Dairy products	Fruits and Vege-tables	other foods
expenditure elasticities	0.698 ***	1.165 ***	1.209 ***	0.908 ***	1.243 ***	0.907 ***
uncompensated price elasticities						
Bread	-0.713 ***	-0.041 ***	-0.161 ***	-0.030 ***	-0.116 ***	-0.075 ***
Meat	0.048 ***	-0.963 ***	-0.058 *	0.105 ***	-0.098 ***	-0.090 ***
Fish	-0.009 *	-0.009 **	-0.963 ***	0.016 ***	-0.014 ***	0.017 ***
Dairy products	0.010	0.034 ***	0.023	-0.935 ***	-0.066 ***	-0.057 ***
Fruits and Vegetables	-0.006	-0.085 ***	-0.077 **	-0.010	-0.863 ***	-0.040 ***
other foods	-0.041 ***	-0.093 ***	0.036	-0.058 ***	-0.074 ***	-0.872 ***
compensated price elasticities						
Bread	-0.580 ***	0.181 ***	0.069 ***	0.143 ***	0.121 ***	0.098 ***
Meat	0.211 ***	-0.690 ***	0.226 ***	0.317 ***	0.194 ***	0.122 ***
Fish	0.015 ***	0.030 ***	-0.922 ***	0.046 ***	0.027 ***	0.047 ***
Dairy products	0.127 ***	0.229 ***	0.226 ***	-0.783 ***	0.142 ***	0.095 ***
Fruits and Vegetables	0.144 ***	0.166 ***	0.183 ***	0.185 ***	-0.596 ***	0.156 ***
other foods	0.070 ***	0.092 ***	0.228 ***	0.087 ***	0.123 ***	-0.728 ***

*** indicates significance at p<0.01; ** indicates significance at p<0.05; * indicates significance at p<0.10

Table A 9: Expenditure elasticities and price elasticities of fish commodity (estimation with quality adjusted prices and consistent two step estimation)

	fresh fish	frozen fish	fresh fish fillets	frozen fish fillets	fish or fish fillets without separation fresh or frozen	breaded fish
expenditure elasticities	0.997 ***	0.999 ***	0.795 ***	0.959 ***	0.665 ***	0.193 **
uncompensated price elasticities						
fresh fish	-1.860 ***	1.413 ***	1.122 ***	1.452 ***	1.135 ***	-0.886 ***
frozen fish	0.473 *	-3.345 ***	0.990 ***	0.975 ***	1.105 ***	1.363 ***
fresh fish fillets	0.792 ***	1.257 ***	-2.138 ***	1.096 ***	0.918 ***	-0.439 ***
frozen fish fillets	0.901 ***	1.690 ***	1.101 ***	-2.020 ***	1.098 ***	-0.362 *
fish or fish fillets without separation fresh or frozen	0.981 ***	0.834 ***	1.054 ***	1.083 ***	-1.764 ***	-0.416 **
breaded fish	-2.288 ***	-2.849 **	-3.159 ***	-3.591 ***	-3.540 ***	-4.702 ***
compensated price elasticities						
fresh fish	-1.772 ***	1.501 ***	1.192 ***	1.536 ***	1.194 ***	-0.869 ***
frozen fish	0.500 *	-3.318 ***	1.012 ***	1.001 ***	1.123 ***	1.368 ***
fresh fish fillets	0.879 ***	1.345 ***	-2.068 ***	1.180 ***	0.976 ***	-0.422 ***
frozen fish fillets	0.956 ***	1.746 ***	1.145 ***	-1.967 ***	1.135 ***	-0.352 *
fish or fish fillets without separation fresh or frozen	1.130 ***	0.983 ***	1.173 ***	1.226 ***	-1.664 ***	-0.387 **
breaded fish	-2.194 ***	-2.755 **	-3.084 ***	-3.501 ***	-3.478 ***	-4.684 ***

*** indicates significance at p<0.01; ** indicates significance at p<0.05; * indicates significance at p<0.10

Table A 10: Expenditure elasticities and price elasticities of fish commodities (estimation without quality adjusted prices and consistent two step estimation)

	fresh fish	frozen fish	fresh fish fillets	frozen fish fillets	fish or fish fillets without separation fresh or frozen	breaded fish
expenditure elasticities	3.884 ***	-0.851	1.142 ***	-2.793 ***	0.599 **	-0.539 ***
uncompensated price elasticities						
fresh fish	-1.588 ***	0.263	-0.032	0.817 ***	0.153 ***	0.249 **
frozen fish	-0.248 **	-1.198	0.050	0.598 **	0.036	-0.017
fresh fish fillets	-0.661 ***	0.615	-1.052 ***	0.966 ***	0.173 ***	0.129
frozen fish fillets	-0.451 ***	0.950 *	0.040	-0.331	0.106 **	-0.135
fish or fish fillets without separation fresh or frozen	-1.036 ***	0.775 **	0.083	1.626 ***	-0.832 ***	0.282 *
breaded fish	0.424 ***	-0.763 *	-0.215	-1.311 ***	-0.279 **	-0.882 ***
compensated price elasticities						
fresh fish	-1.245 ***	0.188	0.069	0.570 ***	0.206 ***	0.202 **
frozen fish	-0.141	-1.221	0.081	0.521 *	0.052	-0.032
fresh fish fillets	-0.322 **	0.541	-0.952 ***	0.722 ***	0.225 ***	0.082
frozen fish fillets	-0.236 *	0.903	0.103	-0.485	0.139 **	-0.165
fish or fish fillets without separation fresh or frozen	-0.456 ***	0.648 **	0.253 *	1.209 ***	-0.743 ***	0.202 *
breaded fish	0.789 ***	-0.843 *	-0.108	-1.574 ***	-0.223 **	-0.933 ***

*** indicates significance at p<0.01; ** indicates significance at p<0.05; * indicates significance at p<0.10

Table A 11: Expenditure elasticities and price elasticities of fish commodities (estimation without quality adjusted prices but with consistent two step estimation)

	fresh fish	frozen fish	fresh fish fillets	frozen fish fillets	fish or fish fillets without separation fresh or frozen	breaded fish
expenditure elasticities	0.832 ***	1.029 ***	0.846 ***	1.082 ***	0.655 ***	0.265 ***
uncompensated price elasticities						
fresh fish	1.087 ***	-0.154	0.597 ***	0.115	0.875 ***	-0.841 ***
frozen fish	0.621 ***	3.943 ***	0.486 *	-0.649 *	1.291 ***	1.676 ***
fresh fish fillets	0.618 ***	-0.392	1.269 ***	-0.014	0.892 ***	-0.930 ***
frozen fish fillets	0.756 ***	-0.751	0.623 ***	3.725 ***	1.265 ***	-0.769 ***
fish or fish fillets without separation fresh or frozen	0.508 ***	-0.277	0.535 ***	0.103	0.107 ***	-0.581 ***
breaded fish	-4.639 ***	-3.361 ***	-4.554 ***	-4.257 ***	-5.530 ***	-4.449 ***
compensated price elasticities						
fresh fish	1.161 ***	-0.063	0.672 ***	0.210	0.933 ***	-0.818 ***
frozen fish	0.644 ***	3.972 ***	0.509 **	-0.619 *	1.309 ***	1.684 ***
fresh fish fillets	0.691 ***	-0.303	1.343 ***	0.080	0.949 ***	-0.907 ***
frozen fish fillets	0.802 ***	-0.694	0.670 ***	3.785 ***	1.302 ***	-0.754 ***
fish or fish fillets without separation fresh or frozen	0.632 ***	-0.124	0.661 ***	0.265	0.205	-0.541 ***
breaded fish	-4.561 ***	-3.264 ***	-4.474 ***	-4.155 ***	-5.468 ***	-4.424 ***

*** indicates significance at p<0.01; ** indicates significance at p<0.05; * indicates significance at p<0.10

Table A 12: Expenditure elasticities and price elasticities of fish commodity groups (estimation with quality adjusted prices but without consistent two step estimation)

	fresh fish	frozen fish	fresh fish fillets	frozen fish fillets	fish or fish fillets without separation fresh or frozen	breaded fish
expenditure elasticities	1.218 ***	1.033 ***	1.920 ***	1.179 ***	1.264 ***	1.834 ***
uncompensated price elasticities						
fresh fish	-2.291 ***	0.138 ***	1.222 ***	0.143 **	0.225 ***	-0.256 ***
frozen fish	0.179 ***	-1.926 ***	1.202 ***	0.429 ***	0.120 ***	0.203 ***
fresh fish fillets	0.236 ***	0.169 ***	-8.473 ***	0.252 ***	0.265 ***	-0.055
frozen fish fillets	0.197 ***	0.312 ***	1.471 ***	-2.408 ***	0.232 ***	0.137 **
fish or fish fillets without separation fresh or frozen	0.322 ***	0.110 ***	1.823 ***	0.244 ***	-2.171 ***	-0.232 ***
breaded fish	0.123 ***	0.163 ***	0.766 ***	0.149 **	0.045	-1.392 ***
compensated price elasticities						
fresh fish	-2.059 ***	0.334 ***	1.588 ***	0.368 ***	0.465 ***	0.093
frozen fish	0.465 ***	-1.684 ***	1.652 ***	0.705 ***	0.416 ***	0.633 ***
fresh fish fillets	0.277 ***	0.203 ***	-8.409 ***	0.291 ***	0.307 ***	0.006
frozen fish fillets	0.401 ***	0.485 ***	1.793 ***	-2.211 ***	0.444 ***	0.444 ***
fish or fish fillets without separation fresh or frozen	0.584 ***	0.332 ***	2.236 ***	0.497 ***	-1.898 ***	0.163 ***
breaded fish	0.316 ***	0.327 ***	1.071 ***	0.336 ***	0.246 ***	-1.100 ***

*** indicates significance at p<0.01; ** indicates significance at p<0.05; * indicates significance at p<0.10

Table A 13: QUAIDS estimation results of food commodity groups (estimation with quality adjusted prices and consistent two step estimation)

	1	2	3	4	5	6
	Bread	Meat	Fish	Dairy products	Fruits and Vegetables	other foods
α_i	-0.8789 ***	0.3806 ***	0.0802 ***	-0.3631 ***	1.0000 ***	0.7812 ***
	(-5.97)	(9.57)	(11.46)	(-4.37)	(9.13)	(7.64)
β_i	-0.0449 ***	0.0290 ***	0.0004	-0.0134 ***	0.0337 ***	-0.0048
	(-12.07)	(6.83)	(0.29)	(-5.05)	(9.05)	(-1.29)
γ_{i1}	0.0830 ***	-0.0149 ***	-0.0052 ***	0.0039	-0.0419 ***	-0.0248 ***
	(13.27)	(-5.32)	(-4.56)	(1.51)	(-9.49)	(-7.99)
γ_{i2}		0.0275 ***	-0.0011	0.0114 ***	-0.0047	-0.0182 ***
		(6.6)	(-0.89)	(4.94)	(-1.62)	(-7.47)
γ_{i3}			0.0054 ***	0.0025 ***	-0.0014	-0.0002
			(6.03)	(2.64)	(-1.11)	(-0.23)
γ_{i4}				0.0141 ***	-0.0166 ***	-0.0153 ***
				(5.91)	(-6.48)	(-8.07)
γ_{i5}					0.0702 ***	-0.0056 **
					(13.74)	(-2.05)
γ_{i6}						0.0144 **
						(2.51)
λ_i	-0.0021 ***	0.0008 **	0.0001	-0.0002	0.0016 ***	0.0002
	(-9.05)	(2.26)	(1.14)	(-1.23)	(6.34)	(0.75)
θ_i	0.0147 ***	0.0025 *	-0.0064 ***	0.0058 ***	-0.0263 ***	0.0096 ***
	(-15.12)	(1.87)	(-14.62)	(7.12)	(-22.96)	(10.41)
σ_i	4.1859 ***	-0.7013 ***	-0.0695 ***	1.9778 ***	-2.5349 ***	-2.8581 ***
	(8.31)	(-5.85)	(-5.49)	(6.82)	(-6.72)	(-7.68)

*** indicates significance at $p<0.01$; ** indicates significance at $p<0.05$; * indicates significance at $p<0.10$
z-statistics in parantheses

Table A 14: QUAIDS estimation results of food commodity groups (estimation without quality adjusted prices and consistent two step estimation)

	1	2	3	4	5	6
	Bread	Meat	Fish	Dairy products	Fruits and Vegetables	other foods
a_i	0.2991 ***	0.0975 ***	0.0329 ***	0.1888 ***	0.1970 ***	0.1847 ***
	(35.96)	(8.81)	(8.97)	(28.57)	(21.98)	(28.07)
β_i	-0.0458 ***	0.0362 ***	-0.0039 **	-0.0214 ***	0.0478 ***	-0.0128 ***
	(-12.8)	(6.93)	(-2.38)	(-7.63)	(12.39)	(-4.52)
γ_{i1}	0.0407 ***	0.0016	-0.0084 ***	-0.0097 ***	-0.0111 ***	-0.0132 ***
	(16.55)	(0.74)	(-8.36)	(-6.21)	(-6.03)	(-8.66)
γ_{i2}		0.0224 ***	-0.0038 ***	0.0211 ***	-0.0215 ***	-0.0199 ***
		(5.7)	(-3.12)	(10.54)	(-8.86)	(-10.11)
γ_{i3}			0.0167 ***	0.0000	-0.0051 ***	0.0005
			(14.59)	(0.04)	(-5.14)	(0.57)
γ_{i4}				0.0062 ***	-0.0047 ***	-0.0130 ***
				(3.3)	(-2.91)	(-10.02)
γ_{i5}					0.0485 ***	-0.0062 ***
					(16.4)	(-3.88)
γ_{i6}						0.0254 ***
						(7.5)
λ_i	-0.0022 ***	0.0014 **	-0.0001	-0.0007 ***	0.0021 ***	0.0005 *
	(-8.1)	(2.55)	(-0.8)	(-4.05)	(8.57)	(3.57)
θ_i	0.0135 ***	0.0023 *	-0.0024 ***	0.0062 ***	-0.0246 ***	0.0051 ***
	(15.17)	(1.84)	(-5.68)	(7.93)	(-22.38)	(6.11)

*** indicates significance at p<0.01; ** indicates significance at p<0.05; * indicates significance at p<0.10

z-statistics in parantheses

Table A 15: QUAIDS estimation results of food commodity groups (estimation without quality adjusted prices but with consistent two step estimation)

	1	2	3	4	5	6
	Bread	Meat	Fish	Dairy products	Fruits and Vegetables	other foods
α_i	-0.3608 ***	0.1939 ***	0.1324 ***	-0.2987 ***	0.6630 ***	0.6701 ***
	(-3.74)	(4.78)	(15.92)	(-4.36)	(7.45)	(10.38)
β_i	-0.0538 ***	0.0420 ***	-0.0179 ***	-0.0233 ***	0.0490 ***	0.0041
	(-14.7)	(9.19)	(-8.4)	(-7.38)	(11.76)	(1.09)
γ_{i1}	0.0718 ***	-0.0102 ***	-0.0056 ***	0.0022	-0.0339 ***	-0.0244 ***
	(14.78)	(-3.22)	(-3.63)	(0.91)	(-8.51)	(-9.78)
γ_{i2}		0.0318 ***	-0.0066 ***	0.0176 ***	-0.0155 ***	-0.0171 ***
		(6.2)	(-4.11)	(6.71)	(-4.75)	(-6.57)
γ_{i3}			0.0291 ***	0.0025	-0.0106 ***	-0.0088 ***
			(19.77)	(1.84)	(-6.75)	(-6.43)
γ_{i4}				0.0150 ***	-0.0176 ***	-0.0197 ***
				(5.45)	(-6.63)	(-9.87)
γ_{i5}					0.0765 ***	0.0010
					(15.07)	(0.41)
γ_{i6}						0.0202
λ_i	-0.0024 ***	0.0013 ***	-0.0007 ***	-0.0005 **	0.0021 ***	-0.0001
	(-11.85)	(4.11)	(-6.22)	(-1.97)	(6.87)	(-0.63)
θ_i	0.0173 ***	0.0000	-0.0038 ***	0.0086 ***	-0.0295 ***	0.0074 ***
	(16.41)	(-0.02)	(-7.51)	(9.44)	(-23.12)	(7.37)
σ_i	2.4614 ***	-0.2233 *	-0.1564 ***	1.7882 ***	-1.3905 ***	-2.4793 ***
	(7.41)	(-1.86)	(-12.44)	(7.55)	(-4.52)	(-10.62)

*** indicates significance at p<0.01; ** indicates significance at p<0.05; * indicates significance at p<0.10
z-statistics in parantheses

Table A 16: QUAIDS estimation results of food commodity groups (estimation with quality adjusted prices but without consistent two step estimation)

	1	2	3	4	5	6
	Bread	Meat	Fish	Dairy products	Fruits and Vegetables	other foods
α_i	0.2910 ***	0.1223 ***	0.0287 ***	0.1791 ***	0.1894 ***	0.1895 ***
	(41.22)	(12.28)	(8.23)	(30.15)	(23.19)	(30.76)
β_i	-0.0432 ***	0.0309 ***	0.0049 ***	-0.0127 ***	0.0401 ***	-0.0200 ***
	(-13.4)	(6.83)	(3.06)	(-4.64)	(10.86)	(-7.05)
γ_{i1}	0.0397 ***	0.0005	-0.0036 ***	-0.0090 ***	-0.0113 ***	-0.0163 ***
	(18.63)	(0.29)	(-4.39)	(-7.1)	(-6.55)	(-12.36)
γ_{i2}		0.0144 ***	-0.0009	0.0153 ***	-0.0133 ***	-0.0161 ***
		(4.98)	(-0.96)	(10.25)	(-6.58)	(-10.58)
γ_{i3}			0.0015 **	0.0021 ***	-0.0013	0.0022 ***
			(2.32)	(3.05)	(-1.5)	(3.14)
γ_{i4}				0.0079 ***	-0.0043 ***	-0.0120 ***
				(5.35)	(-3.11)	(-10.95)
γ_{i5}					0.0385 ***	-0.0083 ***
					(15.1)	(-5.68)
γ_{i6}						0.0178 ***
						(6.46)
λ_i	-0.0022 ***	0.0012 **	0.0003 ***	-0.0004 *	0.0019 ***	0.0008 ***
	(-6.68)	(2.61)	(2.06)	(-1.5)	(5.05)	(2.77)
θ_i	0.0116 ***	0.0049 ***	-0.0042 ***	0.0038 ***	-0.0218 ***	0.0057 ***
	(14.06)	(4.27)	(-10.28)	(5.39)	(-22.71)	(7.77)

*** indicates significance at $p<0.01$; ** indicates significance at $p<0.05$; * indicates significance at $p<0.10$
z-statistics in parantheses

Table A 17: QUAIDS estimation results of fish commodities (estimation with quality adjusted prices and consistent two step estimation)

	1	2	3	4	5	6
	fresh fish	frozen fish	fresh fish fillets	frozen fish fillets	fish or fish fillets without separation fresh or frozen	breaded fish
α_i	1.5364 ***	2.7153 ***	1.0081 ***	1.8041 ***	0.6229 ***	-6.6868 ***
	(-13.75)	(-7.41)	(7.93)	(8.05)	(4.55)	(-13.01)
β_i	0.0243 ***	0.0081	-0.0043	0.0136	-0.0359 **	-0.0059
	(2.73)	(1.04)	(-0.42)	(1.6)	(-2.29)	(-0.16)
γ_{i1}	-0.1384 ***	0.0754 ***	0.1271 ***	0.1452 ***	0.1562 ***	-0.3655 ***
	(-2.75)	(4.64)	(7.97)	(9.68)	(11.77)	(-8.46)
γ_{i2}		-0.1246 **	0.0667 ***	0.0899 ***	0.0436 ***	-0.1510 ***
		(-2.03)	(5.3)	(4.34)	(4.48)	(-3.22)
γ_{i3}			-0.2180 ***	0.1093 ***	0.1417 ***	-0.2268 ***
			(-4.82)	(8.68)	(13.24)	(-5.35)
γ_{i4}				-0.1145 *	0.1083 ***	-0.3381 ***
				(-1.77)	(9.39)	(-5.98)
γ_{i5}					-0.2679 ***	-0.1819 ***
					(-6.62)	(-3.88)
γ_{i6}						0.5323 ***
						(5.00)
λ_i	0.0030 ***	0.0010 **	0.0035 ***	0.0022 ***	0.0063 ***	0.0160 ***
	(3.86)	(2.23)	(4.13)	(3.96)	(4.3)	(4.5)
θ_i	-0.0340 ***	-0.0136 ***	-0.0263 ***	-0.0148 **	-0.0392 ***	0.1278 ***
	(-5.56)	(-3.03)	(-4.18)	(-2.46)	(-5.19)	(17.32)
σ_i	-1.4338 ***	-3.3778 ***	-0.9179 ***	-2.2671 ***	-0.2218	8.2186 ***
	(-8.33)	(-7.37)	(-5.3)	(-8.26)	(-1.16)	(13.15)

*** indicates significance at p<0.01; ** indicates significance at p<0.05; * indicates significance at p<0.10
z-statistics in parantheses

Table A 18: QUAIDS estimation results of fish commodities (estimation without quality adjusted prices and consistent two step estimation)

	1	2	3	4	5	6
	fresh fish	frozen fish	fresh fish fillets	frozen fish fillets	fish or fish fillets without separation fresh or frozen	breaded fish
α_i	0.3435 *** (19.49)	0.0478 *** (4.87)	0.2355 *** (14.97)	0.0568 *** (4.46)	0.3713 *** (21)	-0.0550 *** (-3.08)
β_i	0.1495 *** (8.99)	-0.0217 *** (-2.7)	0.0182 (1.34)	-0.0923 *** (-10.16)	-0.0079 (-0.48)	-0.0457 *** (-4.17)
γ_{i1}	0.0256 (0.95)	-0.0071 (-0.85)	0.0025 (0.24)	-0.0143 (-1.46)	0.0084 (0.94)	-0.0151 (-1.54)
γ_{i2}		-0.0085 (-0.21)	0.0049 (0.51)	0.0203 (1.32)	0.0013 (0.2)	-0.0108 (-0.93)
γ_{i3}			-0.0014 (-0.05)	0.0038 (0.32)	0.0120 (1.32)	-0.0218 * (-1.89)
γ_{i4}				0.0134 (0.39)	0.0074 (0.85)	-0.0305 ** (-2.42)
γ_{i5}					0.0014 (0.06)	-0.0305 *** (-3.07)
γ_{i6}						0.0785 *** (3.28)
λ_i	-0.0305 *** (-5.27)	0.0084 *** (2.94)	0.0017 (0.35)	0.0341 *** (10.63)	0.0151 *** (2.56)	0.0288 *** (6.02)
θ_i	-0.0178 *** (-5.04)	-0.0015 (-0.68)	-0.0150 *** (-4.15)	0.0013 (0.42)	-0.0229 *** (-5.14)	0.0559 *** (14.52)

*** indicates significance at p<0.01; ** indicates significance at p<0.05; * indicates significance at p<0.10

z-statistics in parantheses

Table A 19: QUAIDS estimation results of fish commodities (estimation without quality adjusted prices but with consistent two step estimation)

	1	2	3	4	5	6
	fresh fish	frozen fish	fresh fish fillets	frozen fish fillets	fish or fish fillets without separation fresh or frozen	breaded fish
α_i	1.7754 ***	3.8358 ***	1.7531 ***	3.2838 ***	1.1179 ***	-10.7660 ***
	(16.18)	(6.93)	(16.26)	(15.34)	(10.48)	(-17.76)
β_i	0.0026	0.0062	0.0012	0.0211 ***	-0.0347 ***	0.0036
	(0.52)	(0.82)	(0.2)	(3.43)	(-4.75)	(0.23)
γ_{i1}	0.2816 ***	-0.0049	0.0452 **	0.0295 *	0.0443 ***	-0.3957 ***
	(5.15)	(-0.32)	(2.56)	(1.82)	(2.99)	(-8.9)
γ_{i2}		0.2687 ***	-0.0177	-0.0341	-0.0132	-0.1988 **
		(3.02)	(-1.11)	(-1.3)	(-1.14)	(-2.26)
γ_{i3}			0.3127 ***	0.0157	0.0513 ***	-0.4072 ***
			(5.12)	(0.84)	(3.45)	(-8.06)
γ_{i4}				0.5226 ***	0.0203	-0.5540 ***
				(6.66)	(1.43)	(-8.07)
γ_{i5}					0.1677 ***	-0.2703 ***
					(3.6)	(-6.35)
γ_{i6}						1.0347 ***
						(8.13)
λ_i	0.0025 ***	0.0004 **	0.0021 ***	0.0010 ***	0.0045 ***	0.0105 ***
	(7.6)	(2.28)	(5.98)	(4.58)	(7.58)	(8.57)
θ_i	-0.0324 ***	-0.0174 ***	-0.0289 ***	-0.0261 ***	-0.0395 ***	0.1444 ***
	(-5.11)	(-3.4)	(-4.49)	(-4.25)	(-5.13)	(19.2)
σ_i	-2.0745 ***	-4.7898 ***	-2.0968 ***	-4.0454 ***	-0.9666 ***	13.9732 ***
	(-13.51)	(-6.86)	(-13.91)	(-15.24)	(-6.65)	(18.24)

*** indicates significance at $p<0.01$; ** indicates significance at $p<0.05$; * indicates significance at $p<0.10$
z-statistics in parantheses

Table A 20: QUAIDS estimation results of fish commodities (estimation with quality adjusted prices but without consistent two step estimation)

	1	2	3	4	5	6
	fresh fish	frozen fish	fresh fish fillets	frozen fish fillets	fish or fish fillets without separation fresh or frozen	breaded fish
α_i	0.4581 ***	0.0579 ***	0.2742 ***	0.0134	0.3499 ***	-0.1536 ***
	(25.23)	(6.47)	(20.44)	(1.01)	(20.14)	(-9.72)
β_i	0.1209 ***	-0.0158 **	0.0076	-0.0653 ***	-0.0301 *	-0.0172 *
	(7.83)	(-2.08)	(0.62)	(-6.78)	(-1.92)	(-1.73)
γ_{i1}	-0.2416 ***	0.0379 ***	0.0545 ***	0.0463 ***	0.0790 ***	0.0239 **
	(-10.7)	(4.13)	(5.99)	(5.05)	(10.83)	(2.49)
γ_{i2}		-0.2170 ***	0.0416 ***	0.0721 ***	0.0278 ***	0.0376 ***
		(-6.41)	(6.03)	(6.22)	(5.31)	(3.66)
γ_{i3}			-0.2425 ***	0.0502 ***	0.0715 ***	0.0246 ***
			(-12.33)	(6.54)	(11.33)	(2.82)
γ_{i4}				-0.2400 ***	0.0488 ***	0.0226 *
				(-9.39)	(7.16)	(2.2)
γ_{i5}					-0.2340 ***	0.0069
					(-13.59)	(0.88)
γ_{i6}						-0.0677 ***
						(-3.23)
λ_i	-0.0205 ***	0.0061 **	0.0060	0.0246 ***	0.0225 ***	0.0387 ***
	(-3.97)	(2.26)	(1.43)	(7.47)	(4.13)	(9.15)
θ_i	-0.0175 ***	-0.0017	-0.0154 ***	0.0010	-0.0215 ***	0.0549 ***
	(-5.12)	(-0.77)	(-4.36)	(0.34)	(-4.91)	(14.43)

*** indicates significance at p<0.01; ** indicates significance at p<0.05; * indicates significance at p<0.10

z-statistics in parantheses

7. Simulating the benefits from aquaculture R&D – the impact of elasticities and spillovers

7 Simulating the benefits from aquaculture R&D: the impact of elasticities and spillovers

Abstract

Aquaculture is increasingly important for the future supply of fish because of steadily increasing demand while supply from fisheries is stagnating. Despite the small size of their aquaculture industries some German states have initiated sizeable aquaculture R&D-programs to foster local aquaculture industries. IFPRI´s DREAM model is used to estimate the economic effects of aquaculture R&D conducted in Germany. As the knowledge on parameters like R&D-spillovers, demand, income and supply elasticity is uncertain, several scenarios are run. In one series of the scenarios the size of R&D-spillovers across EU-15-countries is correlated to the strength of fishery and aquaculture research cooperation that have been measured in a bibliometric study. The results of this paper provide important implications for political decisions concerning the allocation of public funds for R&D-projects in aquaculture.

1 Introduction

Animal husbandry is undergoing a rapid revolution. To the small number of economically relevant domesticated terrestrial animal species a large number of aquatic species have been added during the past decades and more will follow soon (DUARTE et al. 2007).

The husbandry of aquatic species is not a recent invention. In China, aquaculture has been practiced for economic gain since 475 B.C. (NASH 2011). Aquaculture has, however, not been an important source of food until recently. In 1950, when world population stood at 2.5 billion people, world aquaculture production had reached 1 million tons, equivalent to 0.4 kg of aquaculture products per capita. Until 2010, world population had grown by 174 percent to 6.9 billion people but aquaculture production had grown more than fifty-fold to 59.9 million tons so that per capita availability of aquaculture products had increased to 8.7 kg in 2010 (FAO 2012a, UN 2012). According to FAO (2011), no other food industry has been growing as quickly as aquaculture.

Two developments have contributed to the rapidly increasing importance of aquaculture as source of food. One is the growing world demand for fish. Global annual per capita consumption of fish has increased from about 10 kg in the 1960s to about 18.6 kg in 2010 (FAO 2011, FAO 2012a). The other reason is the dire state of the world's capture-fishery resources which have been depleted because they are owned by nobody in particular (see Figure 1). World capture fishery production stagnates at around 90 million tons of fish (including finfish, crustaceans and mollusks) per year and it is expected to decline (FAO 2011). Aquaculture production, in contrast, has been growing at about 8.8 percent per year between 1980 and 2010 and it is expected to provide more than half of global fish consumption by 2012 (FAO 2012a).

Figure 1: **World capture and aquaculture production of finfish, 1950-2010 [mio t.]**

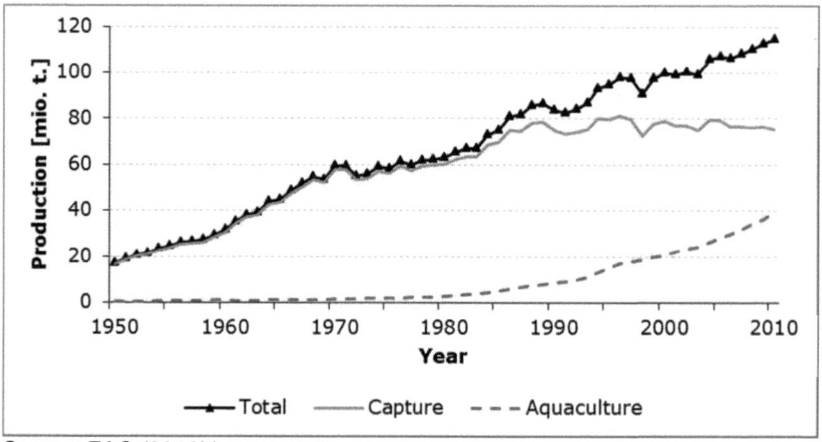

Source: FAO (2012b).

Several factors contribute to the rapid advance of aquaculture. Whereas our traditional farm animals have been domesticated by illiterate savages, aquaculture species are domesticated by highly trained personnel in sophisticated R&D labs. Moreover, many more aquatic than terrestrial animal species are suitable for domestication (DUARTE et al. 2007). In addition, aquaculture production systems do not evolve by trial and error but are designed using knowledge and insights gained in scientific experiments and computer simulations. Finally, aquaculture R&D is, by and large, an open and global undertaking.

This paper is motivated by the belief that economics can contribute to the historical advance of aquaculture. R&D may generate certainly many economic benefits that escape ready measurement. Nevertheless, public support for aquaculture R&D may be strengthened if R&D is informed by an *ex ante* analysis of its potential economic benefits. For this purpose a simulation model is built to assess the potential welfare effects of aquaculture R&D conducted in Germany. The model is distinguished by multiple features: Because aquaculture in Germany is small in comparison to other EU countries, R&D-spillover effects to other EU countries are taken into account. Moreover, the size of R&D-spillovers across EU countries is correlated to the strength of fishery and aquaculture research cooperation that have been measured in a bibliometric study (SEIDEL-LASS 2009). As the knowledge to R&D-spillovers, income and demand elasticity of fish as well as the supply elasticity of aquaculture producers is limited, the sensitivity of consumer and producer benefits is tested to changes in the above named parameters.

This study focuses on the production and consumption of finfish from aquaculture and excludes mollusks, crustaceans and aquatic plants. For reasons of data availability, the model is only concerned with the EU-15 member countries. Moreover, possible effects on markets for substitutes or externalities nor on upstream or downstream markets are not considered.

The paper is organized into six sections. After the introduction some background on aquaculture R&D and on the bibliometric study of international cooperation in fishery research is provided. In section 3 the standard theory of measuring the welfare benefits of R&D is recapitulated and in section 4 the DREAM-simulation model is introduced (ALSTON *et al.* 1995; WOOD *et al.* 2000) together with the data that were used to specify the model. In section 5 the four scenarios are presented together with their model results. Section 6 concludes the paper.

2 Aquaculture Production and R&D-Networks in Germany and in the EU

2.1 Aquaculture Production in Germany and the EU-15

Aquaculture is a small industry in Germany compared to the industries of the major aquaculture producers in the EU. In 2010 Germany produced some

40,000 tons of aquaculture products or 3.5 percent of EU-15 aquaculture production in that year (Table 1). Germany's contribution to world aquaculture production is insignificant at 0.07 percent of the world total.

Had the EU-15 existed in 1970 it would have contributed more than 16 percent to world aquaculture production which stood at 2.56 million tons in that year. Even though aquaculture production in the EU-15 has nearly trebled to 1.17 million tons in 2010, its share in world aquaculture production has dropped to 2 percent because world production has grown more than twentyfold to 59 million tons in 2010.

Table 1: Development of aquaculture production[1] in the EU-15 and in the world, 1970-2010.

	2010			1970			Production growth
	t	% EU-15	% World	t	% EU-15	% World	(p.a. in %)
Spain	252,351	21.6	0.43	156,200	37.4	6.10	1.2
France	224,400	19.2	0.38	106,444	25.5	4.16	1.9
United Kingdom	201,091	17.2	0.34	444	0.1	0.02	16.5
Italy	153,486	13.2	0.26	28,632	6.9	1.12	4.3
Greece	113,486	9.7	0.19	1,040	0.2	0.04	12.4
Netherlands	66,945	5.7	0.11	86,000	20.6	3.36	-0.6
Ireland	46,185	4.0	0.08	3,701	0.9	0.14	6.5
Germany	40,694	3.5	0.07	23,477	5.6	0.92	1.4
Denmark	39,507	3.4	0.07	9,272	2.2	0.36	3.7
Finland	11,772	1.0	0.02	999	0.2	0.04	6.4
Sweden	10,644	0.9	0.02	373	0.1	0.01	8.7
Portugal	3,190	0.3	0.01	47	0.0	0.00	11.1
Austria	2,167	0.2	0.00	870	0.2	0.03	2.3
Belgium	239	0.0	0.00	-	-	-	-
Luxembourg	-	-	-	-	-	-	-
EU-15	1,166,156	100	1.97	417,499	100.0	16.31	2.6
World	59,058,335			2,560,554			8.2

(1) including finfish, crustaceans, mollusks, cephalopods; excluding aquatic plants and mammals

Source: FAO (2012b), own calculations.

In 2010 the five largest EU-15 producers jointly account for 81 percent of total EU-15 production. Germany, which was the (virtual) EU-15 fifth largest aquaculture producer in 1970, has dropped to rank eight even though its aquaculture production has grown by 73 percent in the four decades from 1970 to 2010.

2.2 Aquaculture R&D

Expansion of aquaculture has been driven by consumer demand, better policies and governance, and by R&D breakthroughs (FAO 2011). Even though Europe is a small producer by world standards, Europe's R&D achievements in aquaculture are deemed "remarkable" by FAO (2011, p. 155). ASCHE (2008) states that the most important drivers in the development of modern aquaculture is the control over the production process which allows technological innovations reducing production costs. The prime example is salmon R&D in Norway where production costs in 2007 were only one quarter of the production costs in the mid 1980ies. Due to innovations, e.g. in feeding technologies, disease management and breeding technologies, and productivity growth the production cost in Norwegian salmon aquaculture decreased (ASCHE 2008, GUTTORMSEN et al. 2011). But the EU also has invested heavily in aquaculture R&D. Between 1994 and 1998 the EU spend close to € 60 mio. in the 4th Research Framework Programme and between 1998 and 2002 close to € 88 mio in the 5th Research Framework Programme (EAS 2006). During the 6th Research Framework Programme (2002–2006) aquaculture R&D has attracted close to € 100 mio. and the EU-Commission regards the continued R&D support as essential for the development of aquaculture (EU 2009).

Even though Germany's aquaculture production is currently low, some states in Germany, such as Schleswig-Holstein, a northern seaboard state, have launched sizeable aquaculture R&D projects that are co-funded by the EU. Such projects tend to be justified by a wide range of politically attractive goals and their immediate economic impact on consumers and producers may not be the most important consideration for their promoters and funding agencies. Although local interests may loom large on the agendas of local funding agencies, R&D research issues of general interest are not suppressed, and local funding agencies make no efforts to prevent R&D to spill over to other aquaculture producing states and countries.

2.3 R&D spillovers and networks

Spillovers of useful knowledge from one application domain to another are ubiquitous in agricultural research (ALSTON 2002) and they are present in aquaculture research. For example, salmon R&D conducted in Norway has spilled over into salmon R&D conducted outside Norway and into R&D on other fish species (TVETERÅS and BJØRNDAL 2001, ASCHE 2008). Mediterranean aquaculture producers, in particular, have appropriated some technologies from Norway to boost their production of sea bream and sea bass (SUBASINGHE et al. 2003).

"Spillover" is a metaphor but the term does not specify a mechanism by which useful knowledge actually moves from the "haves" to the "have-nots". Identifying communication channels through which such knowledge may be transferred is one step towards specifying a spillover mechanism. Co-authorships of research papers are such communication channels and are readily measurable with the help of bibliographic databases, such as ISI's Web of Science, and bibliometric methods (GLÄNZEL 2003). In many agricultural R&D studies it is assumed that spillovers are based on geographical proximity or on the similarity of the agricultural production mix (WANG et al. 2012). Knowledge flows between countries have also been approximated by patent citations (SCHERER 1984, JAFFE et al. 1993, MAURSETH and VERSPAGEN 2002) or are based on international trade flows (COE and HELPMAN 1995).

Given the fundamentally unobservable character of knowledge spillovers, directly quantifying their magnitude is a difficult task. To overcome this problem the spillover are related on a network analysis of co-authored publications in aquaculture and fisheries in EU-15-countries. Co-authorships of nearly 13,750 scientific papers published in the aquaculture and fishery research journals that are covered by ISI's Web of Science have been measured and analyzed by SEIDEL-LASS (2009) for the period 1990 to 2005. Based on the publications for 2005, 113 publications for which the author address information indicated residence in an EU-15 member country were selected. Nevertheless, as spillovers are hard to measure, some simulations are run with other spillover-matrices to analyze the sensitivity of the results.

3 Basic economics of R&D impact

For the evaluation of R&D benefits, a standard commodity market model with linear supply and demand based on ALSTON et al. (1995) is used. The formal model is presented in the appendix. R&D is assumed to lead to a parallel downward shift of the supply curve, which is shown in Figure 2. There was a long debate in the agricultural R&D literature on how to best represent the impact of R&D on supply – as a parallel or as pivotal shift of the supply curve. The choice is not trivial because it can significantly influence the magnitude and the distribution of estimated research benefits (ALSTON et al. 1995; ROSE 1980). With parallel supply shifts, producers always benefit from research unless supply is perfectly elastic or demand is perfectly inelastic. In the case of a pivotal shift, in contrast, producers only benefit when demand is elastic (ALSTON et al. 1995). The suggestion by ROSE (1980) is followed in this study and a parallel shift of supply is assumed. This has the additional advantage that one does not need to be concerned with the functional forms of supply and demand for fish (ALSTON et al. 1995).

In Figure 2, S_0 represents the initial supply of the product and the demand curve is given by D. The initial market equilibrium is given by price P_0 and quantity Q_0.

Suppose that R&D results in yield-increasing or input-saving technologies. This can be expressed as a reduction in per unit production costs, k. In the graph, this is expressed as a parallel downward shift of the supply curve from S_0 to S_1. The demand curve D is unaffected by R&D and market equilibrium after the supply shift is given by P_1 and Q_1. Compared to the initial equilibrium (P_0, Q_0) the new equilibrium (P_1, Q_1) is characterized by a higher production and consumption volume, and a lower price.

Figure 2: Surplus distribution in the basic model of research benefits

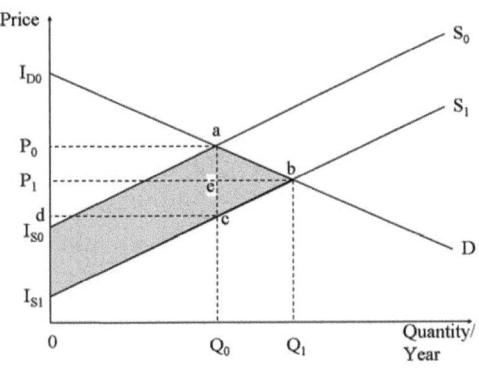

Source: ALSTON et al. (1995).

The producer surplus after the supply shift is equal to the triangle P_1bI_{S1}. The change in producer surplus is shown by the area P_1bI_{S1} minus P_0aI_{S0}. The consumer surplus after the supply shift is equal to the area P_1bI_{D0} and its change corresponds to the area P_0abP_1. The total benefit from the R&D induced supply shift is equal to the shaded area beneath the demand curve D and the supply curves S_0 and S_1 (area $I_{S0}abI_{S1}$). Total benefits can be divided into two

177

parts: The area $I_{S0}acI_{S1}$ is the cost saving on the original quantity Q_0. The area abc is the economic surplus due to the increment in production and consumption.

Spillovers occur if R&D results from one country i are also adopted in another country j. The supply shift in country i at time t, $k_{i,t}$, is then transferred to country j via a spillover coefficient θ_{ji}. The strength of the supply shift $k_{j,t}$ in country j therefore equals $k_{i,t} \times \theta_{ji}$.

The magnitude of the spillover usually ranges between 0 and 1 but may be greater than 1 if the research results are better suited to the country into which the new knowledge or new technology spills than the country where it was done (ALSTON et al. 1995). In the scenario analysis the spillover coefficient θ_{ji} is constrained between 0 and 1, implicitly assuming that, in the best case, the spillover of the research results may lead to equal production cost reductions in the technology-adopting and technology-originating country. In one of the four scenarios θ_{ji} is based on the number of co-authored papers of each country pair identified by the bibliometric analysis and not on the number of citations received from one country. However, it is assumed that the likelihood of transferring knowledge or new technologies from country i to j is higher if researchers from country i and j collaborate in one research project than just referring to a scientific paper in their own publications.

If x_{ji} is the number of collaborations in aquaculture and fisheries research publications between countries i and j, then the spillover coefficient θ_{ji} is calculated by dividing each number by the maximum of the observed number of collaborations (Formula 1):

(1) $\theta_{ji} = \frac{x_{ji}}{\max x_{..}}$.

The spillover coefficients from Germany to the other EU-15 member countries are shown below in Table 3.

4 DREAM and data for its specification and parameterization

DREAM is a software package that implements the graphical model presented above.[7] DREAM has been used in several R&D impact studies (YOU and BOLWIG 2003; BENIN and YOU 2007, JONES *et al.* 2005) including studies of the degree and scope of R&D spillovers (OMAMO *et al.* 2006).

DREAM requires that markets always clear. This is ensured by introducing a virtual country, the "Rest of the World" (ROW), which meets excess demand from the EU-15, which is by far the largest single market for imported fish (FAO 2011).

For each country the market for aquaculture fish has to be specified for the first period $t=0$. The markets are characterized by (i) quantities of supply and demand; (ii) exogenous growth of supply and demand; (iii) elasticities of supply and demand; (iv) initial prices; (v) supply shift parameter $k_{i,t}$, and (vi) technology spillover parameter θ_{ji}.

Data on quantities and values of aquaculture production were obtained from FAO's Fishstat J database (FAO 2012b). Initial market prices were calculated by dividing values by quantities.

Potential growth of aquaculture production depends on a number of factors other than innovation, such as market demand, feed supply, and environmental constraints (FAILLER 2007). The projection of DELGADO *et al.* (2003) includes technological change and changes in investment and results in an estimated annual percentage growth rate of 2.1 percent for EU-15 aquaculture production between 1997 and 2020. FAILLER`S (2007, 2008) prediction is based on past growth rates of EU-15-countries aquaculture sector and he predicts an annual percentage growth rate for EU-15 aquaculture production of less than 0.7 percent for the period 1998 to 2030. The EU-15 aquaculture production of finfish stagnated or often showed a slight decrease in the period 2000 to 2010 (FAO 2012b). It is therefore assumed that innovations enabled by aquaculture R&D are the only source of growth and that the exogenous growth rate for aquaculture supply is zero.

---

[7] The formal model is presented in the appendix.

Data on the consumption of farmed finfish are unavailable. FAOSTAT (FAO 2012c) provides data on the food fish supply which can be equated with the consumption of fish. These data include fish from both capture fisheries and aquaculture. The share of fish from aquaculture increased steadily in the last years and FAO estimates this share to be 29 percent in the year 2010 in the world excluding China (FAO 2012a). As the EU-15 share of total aquaculture production on total production is lower and accounts for 19.7 percent a share of 25 percent is assumed for the model runs.

FAILLER (2007, 2008) predicts that per capita fish consumption will slightly increase until 2030 for most EU-15-countries, with the exception of Ireland, Portugal, Spain, and Sweden. FAILLER'S (2007, 2008) projections on fish consumption are based on national trends but exclude economic factors like income growth. Much of the change in the level and structure of fish consumption reflects more subtle and complex demographic and behavioral variables. Ageing populations, changing gender roles, smaller household sizes, dietary concerns, food safety issues as well as ethical concerns are evident throughout Europe (EUROPEAN COMMISSION 1999). It is nearly impossible to account for all of these factors and the exogenous consumption growth (ΔD_{fish}) is therefore estimated as the sum of the population growth rate (ΔP) and the income growth rate (ΔG) weighted by the income elasticity (η) (OHKAWA 1956 quoted by STEVENS 1963):

(2) $\quad \Delta D_{fish} = \Delta P + \Delta G * \eta.$

Data for population growth rate are taken from UN world population prospects (UN 2012). The income growth rate is based on GDP growth estimated by FOURÉ et al. (2012). To reflect the variety of income elasticity estimations (WESTLUND 2005, ASCHE and BJØRNDAL 2001, BJØRNDAL et al. 1992) the income elasticity of fish is modeled with values of 0.2, 0.6, and 1.0 in this study.

Further, elasticities for demand and supply of finfish from aquaculture have to be quantified. The review of studies on demand elasticities for fish by ASCHE et al. (2005) indicates that demand in most markets is price elastic but for some aquaculture species demand seems to become less elastic with increases in supply. A meta-analysis of price elasticities by GALLET (2009, 2010) showed a median price elasticity of -0.8 for fish. DELGADO *et al.* (2003) suggest that a reasonable range of own price elasticities is between -0.8 and -1.5. Own calculations resulted in price elasticities of fish in the range from -0.25 to -0.96 in Germany (GUETTLER 2013, Chapter 6 in this dissertation). In the model runs demand elasticities (ε^D) of -0.2, -0.8, and -1 are assumed for each EU-15 country.

No studies that report empirical estimates of price elasticities of EU-15 aquaculture supply could be found. DEY *et al.* (2004) estimated aquaculture supply elasticities between 0.28 and 1.24 for some developing Asian countries. ANDERSEN *et al.* (2008) analyzed Norwegian salmon producers and estimate a short-run supply elasticity near zero and a long-run supply elasticity of 1.4. STEEN *et al.* (1997) estimated an intermediate (2 years) supply elasticity of 1 and a long-run (4 years) supply elasticity of 1.54 for Norwegian farmed salmon and this long-run estimation is adopted by KINNUCAN and MYRLAND (2000). BONNIEUX *et al.* (1993) used a short-term supply elasticity of 1.1 and a long-term price elasticity of 2.5 for the modeling of the French trout production sector. Supply in the U.S. of catfish from aquaculture seems to be inelastic with a supply elasticity of 0.15 (ENGLE *et al.* 1998). For lack of better information, supply elasticities (ε^S) of 0.5, 1, and 1.5 are used in the model simulation.

Elasticities of demand and supply play a crucial role in the economic model and for the calculation of research induced benefits. For demand and supply elasticities three different values are used for each country in our scenario analysis. The elasticities are based on published results and some economic thoughts presented subsequently. This parameterization is of course not realistic and the sensitivity of the results to changes in the elasticities will additionally be computed.

The surplus in Figure 2 is composed by the rectangle P_0acd and the triangle *abc*. In the case of total benefits, the rectangle P_0acd is unaffected by the slopes of the supply and demand curves whereas the triangle *abc* is. The more elastic demand or supply is, the larger the triangle and the larger the welfare gain. In the context of estimating research benefits, the triangles are typically very small relative to the rectangles and total benefits are relatively insensitive to elasticities of supply and demand (ALSTON *et al.* 1995).

The distribution of R&D benefits do, however, crucially depend on the price elasticities of supply and demand and the less price-elastic (in absolute terms) market side will be able to appropriate the larger share of research benefits. Only when the price elasticities are of equal absolute magnitudes will the benefits from research be shared equally between producers and consumers (ALSTON *et al.* 1995). In the scenarios of this paper the exogenous growth of demand leads to unequal shares of the total surplus of producers and consumers.

R&D leads to a shift of the supply curve and therefore the supply elasticity is of special importance. OEHMKE and CRAWFORD (2002) showed that the rates of return of a R&D-project can react very sensitive to changes in the parameterization of supply elasticity. In addition, ALSTON *et al.* (1995) state that with an inelastic supply curve, the proportionate cost reduction implied by a proportional rightward shift of supply can be unreasonable, giving rise to overestimated returns. But, if one uses an elastic supply curve, the benefits can be underestimated as well. Therefore, in the absence of better information, a supply elasticity of 1.0 is considered an appropriate starting point (ALSTON *et al.* 1995). To control for the effect of the supply elasticity the scenarios are also modeled with inelastic ($\varepsilon^S=0.5$) and elastic ($\varepsilon^S=1.5$) supply functions.

The impact of R&D on the supply curve has to be parameterized by estimating the R&D-induced reductions of production costs. Production costs of the Norwegian salmon industry decreased by about 6 percent per year between 1986 and 2007 (ASCHE 2008). R&D in salmon aquaculture can be regarded as demanding compared to R&D for other fish species. Advances in the breeding technology have led to a decreasing price of sea bass and sea bream fingerlings by 5.7 percent per year between 1990 and 2003 (BOSTOCK *et al.* 2008). Similar rates of cost reduction may therefore by feasible in EU-15 aquaculture. It is assumed that the new technology leads to a per unit cost reduction of 20 percent, which is modeled as a single fixed shift, implying that without research or without the new technology there would be no shift of the supply curve. If the new technology is adopted by an aquaculture producer in year *t*, the production costs will decrease by 20 percent in that year and will stay at that level for all periods following *t*. Table 2 summarizes the base data used in DREAM simulations.

Table 2: Base data for simulation: EU-15 market for finfish from aquaculture

Country	Supply (1,000 t)	Demand (1,000 t)	Price (1,000 US$/t)	exogenous growth of demand (p.a. in %)		
				(η=0.2)	(η=0.6)	(η=1.0)
Austria	2.2	27.6	12.45	0.34	0.84	1.28
Belgium	0.2	46.5	7.55	0.45	0.88	1.26
Denmark	37.0	22.4	3.64	0.48	0.95	1.36
Finland	11.8	45.9	4.69	0.46	1.01	1.48
France	47.0	340.5	4.68	0.71	1.26	1.74
Germany	35.7	276.1	3.34	-0.06	0.22	0.48
Greece	91.0	41.5	5.18	0.66	1.57	2.30
Ireland	16.9	15.0	5.90	1.20	1.85	2.40
Italy	52.5	228.7	4.97	0.05	0.20	0.34
Luxembourg	0.0	2.3	4.79	1.40	2.09	2.66
Netherlands	6.8	77.9	6.59	0.39	0.85	1.25
Portugal	2.2	134.4	8.33	0.08	0.66	1.16
Spain	59.5	313.0	6.06	0.79	1.54	2.16
Sweden	9.3	55.8	4.35	0.81	1.48	2.04
United Kingdon	169.6	254.4	4.38	0.89	1.59	2.18
*ROW	1340.6	-	4.79	-	-	-

Source: FAO (2012b+c), UN (2012), FOURÉ et al. (2012), own calculations.

5 Scenario Analysis

5.1 Description of the scenarios

For all scenarios the simulation period is 31 years: from 2010 to 2040. This nearly reflects the meta-analysis of ALSTON et al. (2010), who find that in most cases of agricultural R&D the main impact is exhausted within 35 years. Net benefits are discounted to the base year to obtain present values of net benefits. The literature on the choice of discount rates is vast. Like many authors prior to this study, ARROW'S (1995) suggestion is followed and a real discount rate of 3 percent is adopted.

Based on fish-market characteristics for EU-15-countries described previously, four base scenarios are investigated using IFPRI's DREAM model. Measures of producer and consumer surplus are computed and compared between the scenarios.

5.1.1 Scenario 1: R&D effects only in Germany

In the first scenario it is assumed that R&D in Germany induces a reduction of production costs by 20 percent. There are no spillovers from Germany to the rest of the EU-15 (θ_{ji}=0). Furthermore, a five year R&D period (λ_R=5), which is needed to conduct R&D and an adoption lag of ten years (λ_A=10) until the new technology is fully adopted is assumed. These research and adoption lags may be too short compared to actual lags but data on lags for aquaculture technologies are not available. PARDEY and CRAIG (1989) found strong evidence that the impact of agricultural R&D may take as long as thirty years to be felt. ALSTON *et al.* (2008) suggest research and adoption lags of 5 to 10 years or longer in agricultural R&D. This scenario is computed for income elasticities of 0.2, 0.6, and 1, which leads to varying growth rates of fish demand (see Table 2). Additionally, the demand elasticity (ε^D) is set to -0.2, -0.8, and -1 for each country and the supply elasticity (ε^S) is set to 0.5, 1, and 1.5.

5.1.2 Scenario 2: R&D in Germany with spillovers based on the co-authorship network

The second scenario differs from the first in that R&D spillover are taken into account, which are presented in Table 3. Producers in countries like Austria, Belgium, Denmark, Greece, the Netherlands, Portugal and the United Kingdom will benefit from the new technology developed in Germany, while the others will not. For countries where $\theta_{ji}=0$ no co-authored publication could be detected for the year 2005. The strength of the spillover shows the impact of R&D conducted in Germany on the per unit production costs in the spill-in countries. For example, the spillover from Germany to Denmark is 0.25, meaning that 25 percent of the per unit production cost reduction in Germany can be realized by Danish fish farmers. Subsequent to the research lag of 5 years (=λ_R), the new technology can immediately be transferred to and adopted in other spill-in countries. The same adoption curve of the new technology is assumed in each country. As in all scenarios, the sensitivity of the results is tested for changes in the elasticities of income, demand and supply.

Table 3: Spillover coefficients (θ_{ji}) from Germany to EU-15 member countries

	Austria	Belgium	Denmark	Finland	France	Germany	Greece	Ireland	Italy	Netherlands	Portugal	Spain	Sweden	UK
Germany	0.125	0.125	0.25	0	0	-	0.25	0	0	0.25	0.125	0	0	0.375

5.1.3 Scenario 3: R&D in Germany with weak spillovers to all other EU-15-countries

Scenario 3 allows for spillovers which are not based on the co-authorship relationships. Instead it is assumed that spillovers reach all other EU-15 countries with the same strength. In scenario 3 the spillover coefficient is set to 0.1 for each country (θ_{ji}=0.1). All other conditions of scenario 2 stay the same in scenario 3.

5.1.4 Scenario 4: R&D in Germany with strong spillovers to all other EU-15-countries

In contrast to scenario 3 the spillover matrix changes to strong spillovers between Germany and the remaining EU-15 countries. In scenario 4 the spillover coefficient is set to 1 (θ_{ji}=1), meaning that the reduction of the production costs through the new technology is the same for all aquaculture producers in each EU-15 country.

5.2 Results

The estimation of the four scenarios with multiple specifications of the elasticities of demand, income, and supply led to a total of 108 model runs. The detailed results are presented in Table A 1 and in Figure A 1 to Figure A 3 in the appendix. As the research costs were set to zero in all scenarios, the total research benefits could also be interpreted as the upper limit on research investment by the country conducting the research, if the country were prepared to regard the benefits that accrue to other EU-countries as valuable as the benefits that the country is able to reap for itself. Further, the results should be interpreted as rough estimation and the exact results are of secondary importance.

The total net present value (NPV) benefits of the four scenarios range from US $ 409 mio. to US $ 15,505 mio. In scenario 1, German aquaculture producers profit through R&D and gain positive benefits, while all aquaculture producers outside Germany receive a negative net benefit. Additionally, German producers benefit outweighs the negative producer benefits, so that total NPV benefit of producers is positive. Consumers receive positive welfare benefits through slightly reduced prices. The total NPV benefits are lowest in scenario 1 because only the German aquaculture producers benefit from the new technology.

Scenarios 2, 3, and 4 demonstrate the impact of spillovers, either of knowledge or of technologies, to other countries. Spillovers of R&D from Germany lead to large increases in producer surplus in the spill-in countries ($\theta_{ji}>0$). The new technology leads to lower production costs and thus to a higher production and lower prices than it would be the case without research. Spillovers based on the co-authorship relationships (scenario 2) lead to higher benefits as if the spillover is set to 0.1 from Germany to the remaining EU-15 countries (scenario 3). If every aquaculture producer in each EU-15 country benefits from the new technology in the same magnitude ($\theta_{ji}=1$), the benefits for consumers and producers increase dramatically: in scenario 4 the benefits are more than four times higher as in scenario 2. Figure 3 exemplifies these results for the three demand elasticities and an income elasticity of 0.6 and a supply elasticity of 1.

Figure 3: Total net present value benefits of the scenarios with an income elasticity of 0.6 and a supply elasticity of 1 and sensitivity of the results by changes in the demand elasticity [mio. US $]

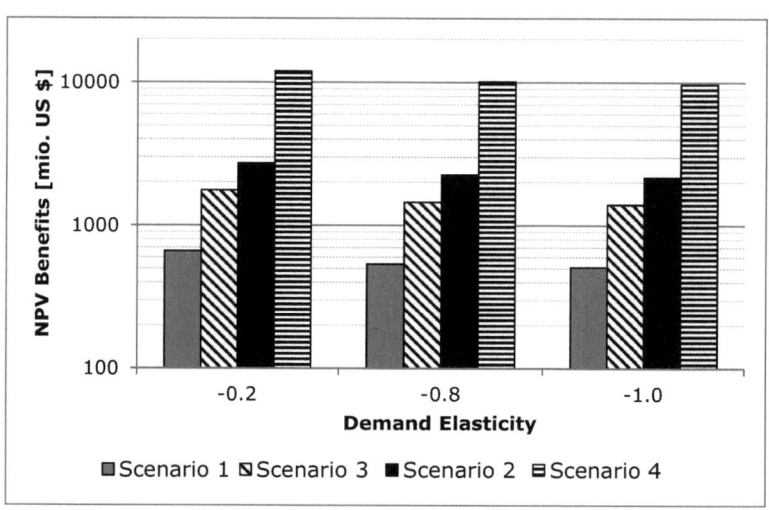

The results also show that when supply becomes more elastic, the total surplus increases in the most cases and the consumers gain a larger share on the total surplus. In scenarios 2, 3 and 4 the total surplus decreases when the supply switches from 1 to 1.5 and when the income elasticity is set to 1 and demand elasticity is -0.2. The same holds for scenario 2 and 3 when the demand elasticity is set to -0.8.

An increasing elasticity of demand leads to a decreasing total surplus and also the consumers' share on total benefits drops. Conversely, producers' share on total benefits increases when demand gets more elastic. The income elasticity has a positive effect on the total NPV benefits, ceteris paribus. However, the effect of the income elasticity on the consumer or producer share of total benefits is ambiguous.

The allocation of the NPV benefits between consumers and producers depends on the scenario settings and fluctuates between 30 percent and 73 percent for the producers and 27 percent to 70 percent for the consumers, respectively. Producers' share is lowest in scenario 1 with an income elasticity of 0.6, a demand elasticity of -0.2 and a supply elasticity of 1.5. The highest producer share evolves in scenario 4 with one-elastic demand and inelastic supply and an income elasticity of 0.2.

5.3 Sensitivity of the results

The knowledge on some model parameters is limited. Therefore, the simulation has been run with different parameter values for the income, supply, and demand elasticity as well as the spillover coefficient. Table 4 gives an overview on the impact of the parameter changes on the result, measured by total surplus. If the demand elasticity is set to -0.2, the total surplus increases on average by 29.5 percent compared to the model runs with a demand elasticity of -1.0. Nearly the same effect results from a change of the supply elasticity from 0.5 to 1.5 (+27.7 percent). A change of the income elasticity from 0.2 to 1.0 leads to a 16 percent higher total surplus on average. Changes in the elasticities may lead to significant changes of the total surplus. However, the effects of changes in the spillover matrix lead to even higher changes of the total surplus. If the spillover coefficient changes from 0 to 1 the total surplus increases by 1,788 percent. The total surplus increases by nearly 590 percent, if the spillover coefficient changes from 0.1 to 1. These results indicate the importance of knowledge spillovers for the surplus of producers and consumers.

Table 4: **Sensitivity of total surplus with respect to changes in the parameter values**

change of parameter from ... to ...	Average change of total surplus [%]
demand elasticity	
from -1.0 to -0.8	5.0
from -1.0 to -0.2	29.5
from -0.8 to -0.2	23.2
income elasticity	
from 0.2 to 0.6	10.2
from 0.2 to 1.0	16.1
from 0.6 to 1.0	5.6
supply elasticity	
from 0.5 to 1.0	17.7
from 0.5 to 1.5	27.7
from 1.0 to 1.5	8.9
spillover coefficient	
from 0 to 0.1	173.8
from 0 to Co-Authorship	324.1
from 0 to 1	1,787.6
from 0.1 to Co-Authorship	54.9
from 0.1 to 1	589.4
from Co-Authorship to 1	345.1

6 Discussion and Closing remarks

This study focuses only on R&D conducted in Germany and its welfare effects on the EU-15. Scenario 1 showed that aquaculture R&D in Germany leads to positive welfare effects in all EU-15-countries, although producers outside Germany receive negative benefits. Scenarios 2, 3, and 4 indicate that international research spillovers significantly increase the benefits from aquaculture R&D. Hence the main qualitative result is that EU support for aquaculture R&D conducted in Germany benefits all spill-in countries of the EU, even when the production of Germany is small.

This study confirms the conclusion of TVETERÅS and BJØRNDAL (2001), who state that public R&D investments have a public good property because all producer countries can benefit from this knowledge or technologies through the effect on the production costs. But not only producers benefit from new technologies, also consumers gain enormous welfare surpluses through declining prices caused by the adoption of the new technology.

This study presented the influence of spillovers, income elasticities, demand elasticities and supply elasticities on the benefits of R&D. The simulation runs with different elasticities showed that the elasticity parameters have an impact on total surplus, but, compared to the effects of spillovers, these impacts are rather small (Table 4). The effects of the spillovers can easily outweigh these effects. Thus it is important that new knowledge or new technologies become known for scientists, researchers and aquaculture producers in each country and that these technologies are adopted in the production sector. The transfer of knowledge can play an important role for producers and consumers but also for researchers, who can use the new knowledge as new input for further research which might lead to further enhancements in the aquaculture technologies.

Results from simulation studies must be interpreted with caution. The data are mostly estimates and some are informed guesses. The results of the scenario analysis are therefore at best rough approximation of actual welfare effects. Much more interesting than the quantitative results are the qualitative insights of our scenario analysis. Spillovers of knowledge lead to an increase of producers and consumers benefits. The dispersion and diffusion of knowledge and research results is therefore an economic activity which should not be neglected and may warrant continued support.

Country-specific elasticities would have made the results more realistic and some countries would have benefited more or less from R&D conducted in Germany. But, the goal of this study is to show the economic effects aquaculture R&D and its spillovers can have on EU-15 in general and not so much to detect the effects for each EU-15-country in detail.

In addition to the usual caveats concerning data availability, functional forms, and other technical matters, there are at least three reasons, where further knowledge is needed to improve the model and its results. The three reasons are: (i) Very little is known about the spillovers from R&D on one fish species to the rest or from one production system to another; (ii) the knowledge about domestic or cross-border adoption lags is less than satisfactory, and finally (iii) the model in this study treats new knowledge gained in R&D only as an output and the fact that such knowledge also is the crucial input for further R&D activities is not taken into account. Outputs of R&D tend, however, to encourage the discovery of even more new knowledge and inventions and a path-dependent, recursive invention process may emerge in aquaculture (ARTHUR 2009).

Aquaculture is a relatively young branch of the food-bioindustries. Like R&D in most young industries, aquaculture R&D has grown rapidly and significant advances can be expected in the near future (STRICKER et al. 2009; FAO 2009). Continued R&D growth and rapid advances will, however, only be realized if investment in aquaculture R&D remain high and commensurate with the benefits that can be had from this exciting branch of food production research.

References

Alston, J.M. (2002): Spillovers. The Australian Journal of Agricultural and Resource Economics, 46(3): 315-346.

Alston, J.M., Norton, G.W. and Pardey, P.G. (1995): Science under scarcity. Cornell University Press, Ithaca and London.

Alston, J.M, Andersen, M.A., James, J.S. and Pardey, P.G. (2010): Persistence pays. U.S. agricultural productivity growth and the benefits from public R&D spending. Springer, New York.

Alston, J.M., Pardey, P.G. and Ruttan, V.W. (2008): Research lags revisited: concepts and evidence from U.S. agriculture, Staff Paper P08-14, InSTePP Paper 08-02, University of Minnesota, December, 2008.

Andersen, T.B., Roll, K.H. and Tveterås, S. (2008): The price responsiveness of salmon supply in the short and long run. Marine Resource Economics 23(4), 425-437.

Arrow, K.J. (1995): Intergenerational equity and the rate of discount in long-term social investment. 11th IEA World Congress, Tunis, December, 1995.

Arthur, W.B. (2009): The nature of technology. Free Press, New York, NY.

Asche, F. (2008): Farming the sea. Marine Resource Economics 23(4): 527-547.

Asche, F. and Bjørndal, T. (2001): Demand elasticities for fish and seafood: a review. Centre for Fisheries Economics, Norwegian School of Economics and Business Administration, Bergen, Norway.

Asche, F., Bjørndal, T. and Gordon, D.V. (2005): Demand structure for fish. SNF Working Paper No. 37/05, Institute for Research in Economics and Business Administration, Bergen, August 2005.

Benin, S. and You, L. (2007): Benefit-cost analysis of Uganda´s clonal coffee replanting program: an ex-ante analysis. IFPRI Discussion Paper 00744, Washington D.C., December, 2007.

Bjørndal, T., Salvanes, K.G. and Andreassen, J.H. (1992): The demand for salmon in France: the effects of marketing and structural change. Applied Economics 24(9): 1027-1034.

Bonnieux, F., Gloaguen, Y., Rainelli, P., Fauré, A., Fauconneua, B., Le Bail, P.-Y., Maisse, G. and Prunet, P. (1993): Potential benefits of biotechnology in aquaculture: The case of growth hormones in French trout farming. Technological Forecasting and Social Change 43: 369-379.

Bostock, J., Muir, J., Young, J., Newton, R. and Paffrath, S. (2008): Prospective analysis of the aquaculture sector in the EU – Part 1: synthesis report. European Commission, Joint Research Centre, Institute for Prospective Technological Studies, Luxembourg.

Coe, D.T. and Helpman, E. (1995): International R&D spillovers. European Economic Review 39(5): 859-887.

Delgado, C.L., Wada, N., Rosegrant, M.W., Meijer, S. and Ahmed, M. (2003): Fish to 2020. International Food Policy Research Institute, World Fish Center, Washington, D.C., Penang.

Dey, M.M., Rodriguez, U.-P., Briones, R.M., Li, C.O., Haque, M.S., Li, L., Kumar, P., Koeshendraja, S., Yew, T.S., Senaratne, A., Nissapa, A., Khiem, N.T. and Ahmed, M. (2004): Disaggregated projections on supply, demand, and trade for developing Asia: Preliminary Results from the ASIAFISH Model. IIFET 2004 Japan Proceedings.

Duarte, C.M., Marbá, N. and Holmer, M. (2007): Rapid domestication of marine species. Science 316: 382-383.

EAS (2006): Impact assessment of the FP4 and FP5 research programme on fisheries, aquaculture and seafood processing research area and the fishery industry. European Aquaculture Society and Oceanic Development, FP6-2003-SSP3-513651. <http://www.easonline.org/component/content/article/1/7-impact-of-eu-research>.

Engle, C.R., Hatch, U., Kinnucan, H., Pomeroy, R., Dellenbarger, L., Dillard, J. and Capps, O. (1998): Analysis of regional and national markets for aquaculture food products in the Southern Region. Southern Regional Aquaculture Center Final Project Report No. 601.

EU (2009): Building a sustainable future for aquaculture: a new impetus for the strategy for sustainable development of European aquaculture. Communication from the Commission to the European Parliament and the Council, COM (2009) 162, Brussels.

European Commission (1999): Forward study of community aquaculture. MacAlister Elliot and Partners Ltd, Lymington.

Failler, P. (2007): Future prospects for fish and fishery products, 4. Fish consumption in the European Union in 2015 and 2030, Part 1, European overview. FAO Fisheries Circular No. 972/4, Part 1, FAO, Rome.

Failler, P. (2008): Future prospects for fish and fishery products, 4. Fish consumption in the European Union in 2015 and 2030, Part 2, Country projections. FAO Fisheries Circular No. 972/4, Part 2, FAO, Rome.

FAO (2009): The state of world fisheries and aquaculture 2008. Food and Agriculture Organization of the United Nations, Rome.

FAO (2011): The state of world fisheries and aquaculture 2010. Food and Agriculture Organization of the United Nations, Rome.

FAO (2012a): The state of world fisheries and aquaculture 2012. Food and Agriculture Organization of the United Nations, Rome.

FAO (2012b): Fisheries and Aquaculture Department, Statistics and Information Service
FishStatJ: Universal software for fishery statistical time series. <http://www.fao.org/fishery/statistics/software/fishstatj/en>.

FAO (2012c): FAOSTAT, <http://faostat3.fao.org/home/index.html>.

Fouré, J., Bénassy-Quéré, A. and Fontagné, L. (2012): The great shift: macroeconomic projections for the world economy at the 2050 horizon. CEPII working paper 2012-03. <http://www.cepii.fr/anglaisgraph/workpap/pdf/2012/wp2012-03.pdf>.

Gallet, C.A. (2009): The demand for fish: a meta-analysis of the own-price elasticity. Aquaculture Economics & Management 13(3): 235-245.

Gallet, C.A. (2010): Meat meets meta: a quantitative review of the price elasticity of meat. American Journal of Agricultural Economics 92(1): 258-272.

Glänzel, W. (2003): Bibliometrics as a research field. Course handouts. <http://citeseerx.ist.psu.edu/viewdoc/download?doi=10.1.1.97.5311&rep=rep1&type=pdf>

Guettler, S. (2013): Demand for fish in Germany. In: Guettler, S.: Studies on the market assessment of aquaculture research by a small producer country, Dissertation, Kiel University, 109-160. <http://eldiss.uni-kiel.de/macau/receive/dissertation_diss_00010683>

Guttormsen, A.G., Ø. Myrland and R. Tveterås (2011): Innovations and structural change in seafood markets and production: special issue introduction. Marine Resource Economics 26(4): 247-253.

Jaffe, A.B., Trajtenberg, M. and Henderson, R. (1993): Geographic localization of knowledge as evidenced by patent citations. Quarterly Journal of Economics 108(3): 577-598.

Jones, R.E., Vere, D.T., Alemseged, Y. and Medd, R.W. (2005): Estimating the economic cost of weeds in Australian annual winter crops. Agricultural Economics 32(3): 253-265.

Kinnucan, H. W. and Myrland, Ø. (2000): Optimal advertising levies with application to the Norway-EU salmon agreement. European Review of Agricultural Economics 27(1): 39–57.

Maurseth, P.B. and Verspagen, B. (2002): Knowledge spillovers in Europe: a patent citation analysis. Scandinavian Journal of Economics 104(4): 531-545.

Nash, C.E. (2011): The history of aquaculture. Wiley-Blackwell, Ames, IO.

Oehmke, J.F. and Crawford, E.W. (2002): The sensitivity of returns to research calculations to supply elasticity. American Journal of Agricultural Economics 84(2): 366-369.

Ohkawa, K. (1956): Economic growth and agriculture. Annals of the Hitotsubachi Academy 7(1): 46-60.

Omamo, S.W., Diao, X., Wood, S., Chamberlin, J., You, L., Benin, S., Wood-Sichra, U. and Tatwangire, A. (2006): Strategic priorities for agricultural development in Eastern and Central Africa. Research Report 150, IFPRI, Washington, D.C.

Pardey, P.G. and Craig, B. (1989): Causal relationships between public sector agricultural research expenditures and output. American Journal of Agricultural Economics 71(1): 9-19.

Rose, R.N. (1980): Supply shifts and research benefits: comment. American Journal of Agricultural Economics 62(4): 834-837

Scherer, F.M. (1984): Using linked patent and R&D data to measure interindustry technology flows. In: Griliches, Z. (ed.): R&D, patents, and productivity. The University of Chicago Press, Chicago and London.

Seidel-Lass, L. (2009): Networks in international aquaculture research: a bibliometric analysis. Cuvillier Verlag, Goettingen.

Steen, F., Asche, F., Salvanes, K.G. (1997): The supply of salmon in EU: a Norwegian aggregated supply curve. SNF Working Paper No. 53/97, Norwegian School of Economic and Business, Bergen.

Stevens, R.D. (1963): The influence of urbanization on the income elasticity of demand for retail food in low income countries. American Journal of Agricultural Economics 45(5): 1495-1499.

Stricker, S., Guettler, S., Schulz, C. and Mueller, R.A.E. (2009): The shape of future aquaculture R&D. Aquaculture Europe 34(2): 18-20.

Subasinghe, R.P., Curry, D., McGladdery, S.E. and Bartley, D. (2003): Recent technological innovations in aquaculture. FAO Fisheries Circular No. 886 (Revision 2), Food and Agriculture Organization of the United Nations, Rome, 59-74.

Tveterås, R. and Bjørndal, T. (2001): Production, Competition and Markets: The evolution of the salmon aquaculture industry. In: Coimbra, J. (ed.) Proceedings of the NATO advanced research workshop on modern aquaculture in the coastal zone - lessons and opportunities, IOS Press, Amsterdam, Oxford, Leipzig, 32-51.

UN (2012): World population prospects: The 2010 revision. United Nations, Department of Economic and Social Affairs, Population Division. <http://esa.un.org/wpp/index.htm>

Wang, S.L., Ball, E., Fulginiti, L.E. and Plastina, A. (2012): Benefits of public R&D in U.S. agriculture: spill-ins, extensions and roads. Selected paper prepared for presentation at the International Association of Agricultural Economists (IAAE) Triennial Conference, Foz do Iguaçu, Brazil, 18-24 August, 2012. <http://ageconsearch.umn.edu/bitstream/126368/2/ICAE_Brazil_new.pdf >

Westlund, L. (2005): Future prospects for fish and fishery products – 5. Forecasting fish consumption and demand analysis: a literature review. FAO Fisheries Circular No. 972/5, Food and Agriculture Organization of the United Nations, Rome.

Wood, S., You, L. and Baitx, W. (2000): DREAM user manual 2000. International Food Policy Research Institute, Washington, D.C.

You, L. and Bolwig, S. (2003): Alternative growth scenarios for Ugandan coffee to 2020. EPTD Discussion Paper No. 98, International Food Policy Research Institute, Washington D.C.

Appendix

Detailed results

Figure A 1: Net present values of all scenarios and sensitivity of the results by changes in the supply elasticity [mio. US $]

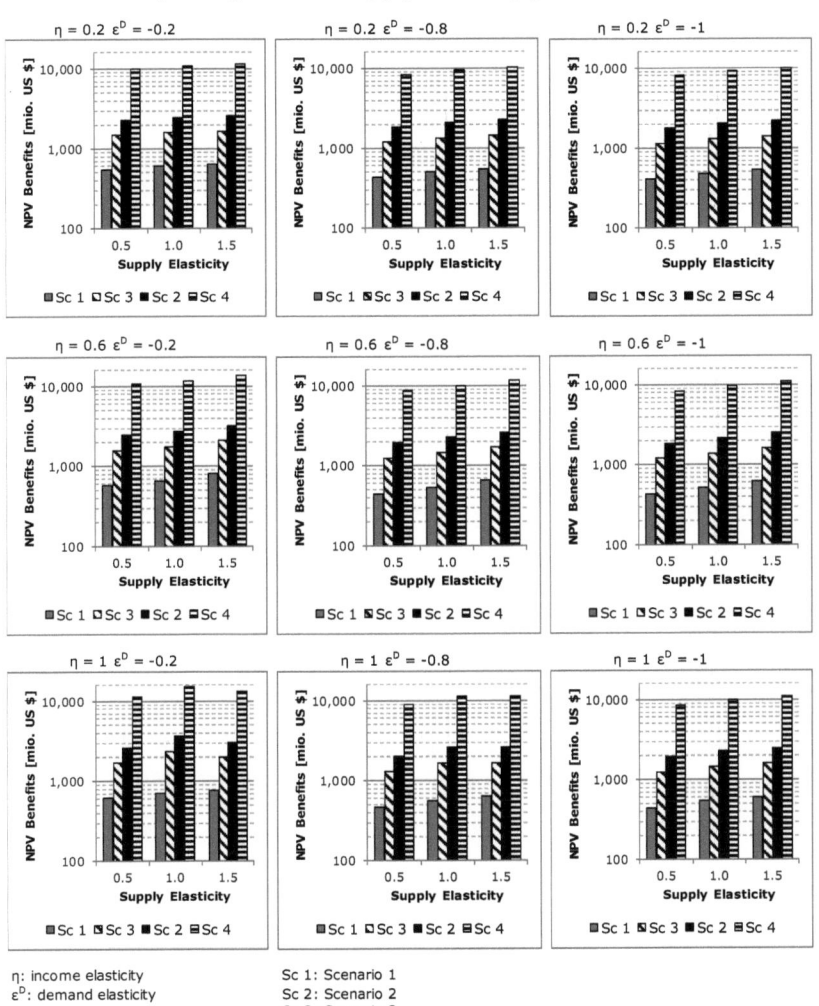

η: income elasticity
ε^D: demand elasticity

Sc 1: Scenario 1
Sc 2: Scenario 2
Sc 3: Scenario 3

Figure A 2: Net present values of all scenarios and sensitivity of the results by changes in the demand elasticity [mio. US $]

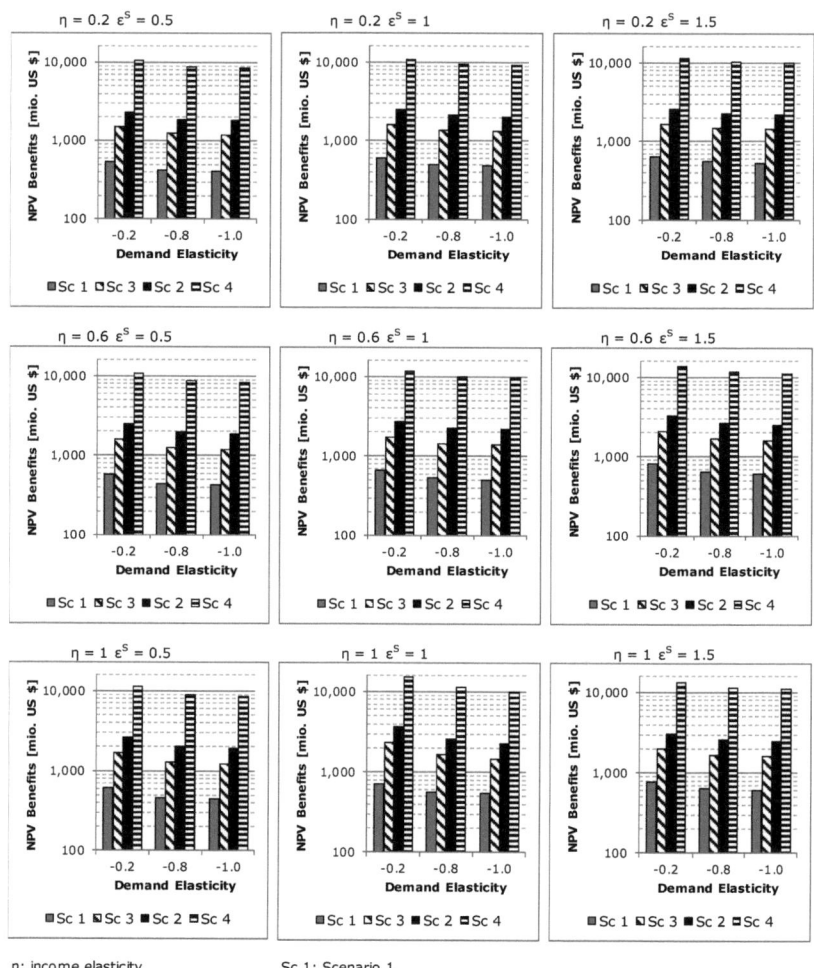

η: income elasticity
ε^S: supply elasticity

Sc 1: Scenario 1
Sc 2: Scenario 2
Sc 3: Scenario 3

Figure A 3: Net present values of all scenarios and sensitivity of the results by changes in the income elasticity [mio. US $]

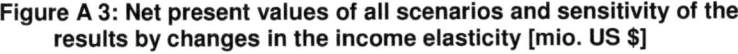

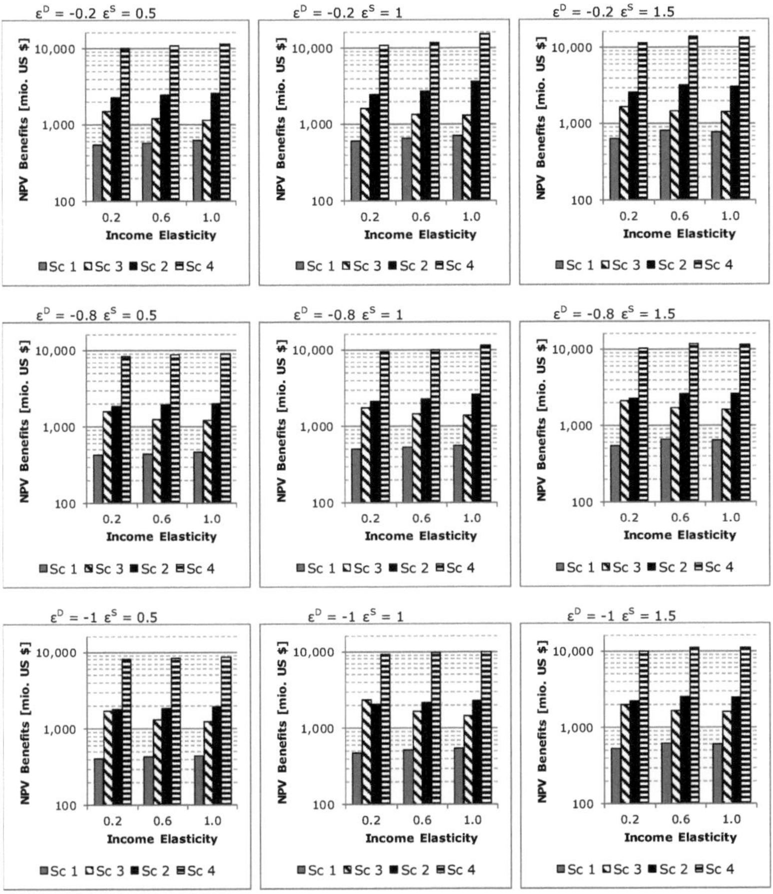

ε^D: demand elasticity
ε^S: supply elasticity

Sc 1: Scenario 1
Sc 2: Scenario 2
Sc 3: Scenario 3

Table A 1: Total net present value benefits of all scenarios [mio. US $] and producers' and consumers' share on total benefits [%]

income elasticity η	demand elasticity ε^D	supply elasticity ε^S	Scenario 1 $\theta_{ji}=0$			Scenario 2 θ_{ji} based on co-authorship			Scenario 3 $\theta_{ji}=0.1$			Scenario 4 $\theta_{ji}=1$		
			Prod. [%]	Cons. [%]	Total [mio. US $]	Prod. [%]	Cons. [%]	Total [mio. US $]	Prod. [%]	Cons. [%]	Total [mio. US $]	Prod. [%]	Cons. [%]	Total [mio. US $]
0.2	-0.2	0.5	43.6	56.4	539	49.7	50.3	2,297	49.7	50.3	1,484	53.4	46.6	10,151
		1.0	39.2	60.8	601	45.0	55.0	2,489	45.0	55.0	1,607	49.3	50.7	10,975
		1.5	38.3	61.7	636	43.5	56.5	2,582	43.4	56.6	1,667	48.1	51.9	11,409
	-0.8	0.5	61.9	38.1	426	67.2	32.8	1,867	67.2	32.8	1,206	70.2	29.8	8,423
		1.0	52.1	47.9	501	57.5	42.5	2,118	57.5	42.5	1,367	61.5	38.5	9,539
		1.5	48.2	51.8	551	53.2	46.8	2,274	53.1	46.9	1,466	57.6	42.4	10,249
	-1.0	0.5	65.7	34.3	409	70.6	29.4	1,801	70.6	29.4	1,163	73.4	26.6	8,159
		1.0	55.3	44.7	481	60.6	39.4	2,046	60.5	39.5	1,320	64.3	35.7	9,255
		1.5	50.8	49.2	532	55.7	44.3	2,205	55.6	44.4	1,421	60.0	40.0	9,987
0.6	-0.2	0.5	44.8	55.2	580	50.0	50.0	2,469	50.1	49.9	1,595	53.2	46.8	10,875
		1.0	39.9	60.1	662	45.2	54.8	2,715	45.1	54.9	1,754	49.1	50.9	11,893
		1.5	30.4	69.6	818	35.1	64.9	3,235	35.1	64.9	2,091	39.3	60.7	13,961
	-0.8	0.5	62.5	37.5	445	67.2	32.8	1,944	67.3	32.7	1,256	70.1	29.9	8,744
		1.0	52.5	47.5	536	57.6	42.4	2,248	57.6	42.4	1,451	61.3	38.7	10,065
		1.5	42.7	57.3	650	47.6	52.4	2,631	47.5	52.5	1,698	51.8	48.2	11,664
	-1.0	0.5	66.1	33.9	425	70.6	29.4	1,866	70.6	29.4	1,205	73.3	26.7	8,428
		1.0	55.6	44.4	512	60.6	39.4	2,159	60.6	39.4	1,393	64.2	35.8	9,714
		1.5	45.9	54.1	617	50.7	49.3	2,511	50.7	49.3	1,619	54.9	45.1	11,202
1.0	-0.2	0.5	45.7	54.3	619	50.3	49.7	2,629	49.6	50.4	1,698	53.1	46.9	11,550
		1.0	40.4	59.6	718	32.7	67.3	3,620	32.7	67.3	2,339	36.0	64.0	15,505
		1.5	39.1	60.9	775	43.6	56.4	3,076	43.6	56.4	1,987	47.8	52.2	13,380
	-0.8	0.5	62.9	37.1	463	67.3	32.7	2,014	67.4	32.6	1,301	70.0	30.0	9,032
		1.0	52.9	47.1	568	50.6	49.4	2,612	50.6	49.4	1,686	54.2	45.8	11,521
		1.5	48.7	51.3	640	53.2	46.8	2,593	53.1	46.9	1,672	57.2	42.8	11,531
	-1.0	0.5	66.4	33.6	440	70.7	29.3	1,925	70.7	29.3	1,243	73.2	26.8	8,669
		1.0	55.9	44.1	539	60.6	39.4	2,262	60.6	39.4	1,460	64.0	36.0	10,133
		1.5	51.2	48.8	611	55.7	44.3	2,489	55.6	44.4	1,605	59.7	40.3	11,132

Model formulation

In this section the formulae of the model described before are presented.

Equation (A1) specifies the supply of fish:

(A1) $Q_{i,t} = \alpha_{i,t} + \beta_i PP_{i,t}.$

The quantity produced Q in country i in the year t is a function of the producer price PP. The slope of the supply curve is determined by β and the axis intercept is given by α.

The quantity consumed in each region is a function of the consumer price in each region $PC_{i,t}$ and the slope of the demand curve is defined by δ, while the intercept of the demand equation is given by γ. Demand for fish C in country i at time t is defined by equation (A2):

(A2) $C_{i,t} = \gamma_{i,t} + \delta_i PC_{i,t}.$

Exogenous growth rates are incorporated to reflect growth in demand and supply that is expected to occur regardless of whether the research program of interest is undertaken.

(A3) $\alpha_{i,t} = \alpha_{i,t-1} + \pi_{i,t}^Q Q_{i,t}$ for $t > 0$

(A4) $\gamma_{i,t} = \gamma_{i,t-1} + \pi_{i,t}^C C_{i,t}$ for $t > 0$

where $\pi_{i,t}^C$ is the exogenous growth rate of demand and $\pi_{i,t}^Q$ is the exogenous growth rate of supply.

The introduction of R&D leads to a downward shift of the supply curve. Let country i undertake a program of research with a probability of success p_i, which, if the research is successful and the results are fully adopted, will yield a cost saving per unit of output equal to c_i percent of the initial price, $PP_{i,0}$ in country i, while a ceiling adoption rate of A_i^{MAX} percent holds in country i. Then it is anticipated that the supply function in region i will shift down (in the price direction) by an amount per unit equal to:

(A5) $k_i^{MAX} = p_i c_i A_i^{MAX} PP_{i,0} \geq 0.$

Our model only considers research lags (λ_R) and adoption lags (years from initial adoption to maximum adoption: λ_A). As disadoption of technologies is not regarded here, the supply shifts (in the price direction) for region i in each year t can be calculated as follows:

(A6) $k_{i,t} = 0$ (for $0 \leq t \leq \lambda_R$)

(A7) $k_{i,t} = k_i^{MAX}(t - \lambda_R)/\lambda_A$ (for $\lambda_R < t \leq \lambda_R + \lambda_A$)

(A8) $k_{i,t} = k_i^{MAX}$ (for $t > \lambda_R + \lambda_A$)

R&D could also lead to supply shifts in other countries, when the new technology is adopted in foreign countries and spillovers occur. These spillover effects of research from one country i to another country j can be parameterized in relation to the supply shifts in region i, whereas θ in equation (A9) is the supply shift in j due to a supply shift in i. This implicitly assumes the same adoption curve in each country.

(A9) $k_{j,t} = \theta_{ji} k_{i,t}$ $\forall\, i,j$

Research effects are included into the supply curve by adjusting the intercept a. In the "with-research" case, denoted by superscript R on the parameters, a is defined by equation (A10):

(A10) $\alpha_{j,t}^R = \alpha_{j,t} + k_{j,t}\beta_{j,t}$

Supply and demand equations in the "with-research" case are given by equation (A11) and (A12), respectively. They reflect the local and spillover effects of research.

(A11) $Q_{i,t}^R = \alpha_{i,t}^R + \beta_{i,t}PP_{i,t}^R$

(A12) $C_{i,t}^R = \gamma_{i,t}^R + \delta_{i,t}PC_{i,t}^R$

The model is solved by introducing a market-clearing rule by equation (A13):

(A13) $Q_t = \sum_{i=1}^n Q_{i,t} = \sum_{i=1}^n C_{i,t} = C_t$

Under the assumption of free trade, producer prices PP equal consumer prices PC in the cases with and without research.

(A14) $PP_{i,t}^R = PC_{i,t}^R = PC_{j,t}^R = PP_{j,t}^R = P_t^R$

(A15) $PP_{i,t} = PC_{i,t} = PC_{j,t} = PP_{j,t} = P_t$

The market clearing prices under free trade are given by equations (A16) and (A17):

(A16) $P_t = (\gamma_t - \alpha_t)/(\beta - \delta)$

(A17) $P_t^R = (\gamma_t - \alpha_t^R)/(\beta - \delta)$

whereas $\gamma_t = \sum_{i=1}^{n} \gamma_{i,t}$; $\alpha_t = \sum_{i=1}^{n} \alpha_{i,t}$; $\alpha_t^R = \sum_{i=1}^{n} \alpha_{i,t}^R$; $\delta_t = \delta = \sum_{i=1}^{n} \delta_{i0} < 0$; and

$\beta_t = \beta = \sum_{i=1}^{n} \beta_{i0} > 0$. As $\gamma_t > \alpha_t^R > \alpha_t$ it follows that $P_t > P_t^R$.

Regional welfare effects through research can be determined and equations (A18) and (A19) show the difference in welfare in the case with research and without research:

(A18) $\Delta PS_{j,t} = \left(k_{j,t} + PP_{j,t}^R - PP_{j,t}\right)\left[Q_{j,t} + 0.5(Q_{j,t}^R - Q_{j,t})\right]$

(A19) $\Delta CS_{j,t} = (PC_{j,t} - PC_{j,t}^R)\left[C_{j,t} + 0.5(C_{j,t}^R - C_{j,t})\right]$,

whereas $\Delta PS_{j,t}$ is the R&D-induced change in producer surplus in region j in year t and $\Delta CS_{j,t}$ is the R&D-induced change in consumer surplus in region j in year t.

For a planning horizon of m years, $\Delta PS_{j,t}$ and $\Delta CS_{j,t}$ can be calculated for each region and each year. After a real discount rate $r_{i,t} = r_{j,t} = r$ is defined, which is the same for each country; it is easy to estimate the present values of benefits (VPS, VCS) through research:

(A20) $VPS_i = \sum_{t=0}^{m} \Delta PS_{i,t}/(1+r)^t$

(A21) $VCS_i = \sum_{t=0}^{m} \Delta CS_{i,t}/(1+r)^t$.

8. General summary, discussion, and outlook

8 General summary, discussion, and outlook

This dissertation consists of six separate essays beginning with an overview of the production and consumption of fish. The remaining essays deal with the current strength and the future direction of aquaculture R&D, international fish trade and the quality of fish trade data, fish demand analysis, and the estimation of economic benefits for producers and consumers induced by aquaculture R&D.

The first essay **"Captured and cultured fish for food"** serves as brief introduction. Compared to agricultural R&D, R&D on aquatic organisms is a relatively young discipline. About 2000 years ago humans had domesticated 90 percent of the land-based species presently cultivated on land. In contrast, about 97 percent of the aquatic species have been domesticated since the start of the 20th century (Duarte et al. 1997). Aquaculture became increasingly important in the last three decades and was responsible for the growth of the total fish production, because global capture production stagnates since the mid-1990s. By volume, China and other Asian countries dominate the aquaculture production. In a global context the aquaculture production of the European Union (EU) is nearly insignificant. The average per capita fish consumption was about 16.4 kg/year in the EU and is below the global average of 18.4 kg/year. The EU-27 is the largest importer of fish and the import values and volumes increased significantly between 2004 and 2009. The last section in this essay analyzes the development of scientific publications. If the number of publications is an indicator for R&D conducted in a specific area, it can be concluded that R&D in aquaculture increased since the beginning of the 1990s. Increasing R&D efforts might lead to a productivity growth in aquaculture so that within a few decades aquaculture fish may be able to substitute for captured fish.

In the second essay **"Testing the quality of international fish trade data"** the quality of trade data of three fish products was tested for the period from 1992 to 2008. The quality of import and export data is important because biased trade data may lead to biased results and delusive conclusions. First, the distributions of the first digits of import and export trade values were tested for their compliance with Benford's law. The results show that the data of some countries deviate significantly from Benford's law. For these countries it might be that irregularities occurred during the process from data collection to publication. However, this does not mean that these countries falsify their data. There is no evidence that countries whose data follow Benford's law are free of errors or manipulations. In particular, the export and import data of Spain followed Benford's law for the three analyzed fish product groups. The export data of the USA and the import data of Germany of the three fish product categories do not follow Benford's law. Overall, no systematic pattern of deviations was detected.

In a second step, bilateral trade data were analyzed. The fact, that transactions are recorded twice in international trade – as export and as import - offers the possibility to compare and judge the reliability of trade data. However, one does not know which value is the true one, if any. The percentage differences of the reported export and import trade values were computed and classified into four cases suggested by Morgenstern (1950). The results of this analysis reveal which country pairs over- or understate their trade values. Some country pairs could be identified whose trade data are always in the same case, e.g. imports are overstated in both countries or exports are understated in both countries, or both. For these country pairs a systematical bias of the trade data can be assumed. If the percentage differences are grouped in more than one case, the likelihood increases that the discrepancies are random. The results indicate that trade data are inaccurate and that large differences between the recorded export and import trade data exist. The more disaggregated the fish trade data are, the larger the share of deviations of more than ±50 percent. The consistency of trade data seems to be higher for highly aggregated product groups and lower for more disaggregated fish product categories. No systematic pattern could be detected so that it is necessary to scrutinize trade data for each specific country pair and commodity. Furthermore, trade data should be inspected before they are used for further analysis and quality programmes of statistical offices should be run to increase the quality and consistency of international trade data.

In the third essay **"Fish in the network – network analysis of international fish trade"** the international trade of cod, shrimps, and salmonidae was analyzed. Methods of the network analysis were applied to get better information on the structure of global trade networks and to characterize the position of individual countries within a network. The most important countries in terms of trade value and number of trading partners were identified. Global network indices were computed for the period from 1990 to 2009. Additionally, the main intermediaries of the three fish species were detected. Results show that international trade with cod, shrimps, and salmonidae increased in terms of value and in terms of countries participating in international fish trade. The number of trade flows for shrimps and salmonidae is significantly higher than for cod indicating that the countries trading with shrimps and salmonidae are much more interlinked. Some network indices indicate that fish exports are based on few countries. But the number of trading partners increases for these export countries. The position of intermediaries in fish trade became less important for cod and salmonidae but became more important for trade with shrimps. The results also reveal differences in the characteristics of the networks for fish from fisheries and from aquaculture.

The fourth essay **"The shape of future aquaculture R&D – results of a Delphi study"** presents the results of an online Delphi study conducted in 2008. More than 270 aquaculture experts were surveyed. In the first round the experts rated the current situation of aquaculture research in high income countries. Moreover, the experts were asked to assess the future development of aquaculture research until the year 2020. Three survey rounds were conducted in order to achieve some convergence of the experts' assessments of the likely future achievements of aquaculture research. Aquaculture experts agree that aquaculture research has achieved much and will continue to do so in the future. They are also convinced that aquaculture research will have a strong impact on the productivity of aquaculture. Fish nutrition, fish breeding, and reproduction are the research areas whose current strength was rated highest. In the year 2020 the expected research achievements will be highest in fish health, fish nutrition, water management, and quality management, according to the experts. Norway was rated as the current and future leading aquaculture research nation by far. Israel and the United Kingdom are expected to be replaced from rank two and three in the future by Spain and the USA. The Delphi study also highlights the current and expected future research fields of different research. The results may help funding agencies and decision makers to identify the promising areas of aquaculture R&D.

The demand for fish was analyzed in the fifth essay **"Demand for fish in Germany"**. Cross sectional household data from the 2003 German income and consumption survey were used to estimate a quadratic almost ideal demand system (QUAIDS). Additionally, price elasticities of fish and fish product categories were estimated. Missing price information and zero observations in the data could lead to biased estimation results so that two methods (consistent-two-step estimation and quality adjusted prices) were applied to control for these effects. However, the sensitivity of the results was also tested to modifications in the estimation procedure with and without these two methods. The estimation with quality adjusted prices and the consistent-two-step estimation showed that the own-price elasticity of fish is about -0.88. Modifications in the estimation method showed that the own-price elasticities of fish range from -0.25 to -0.96. However, the qualitative results remain unchanged and demand for fish in Germany is inelastic. Cross-price elasticities of fish and other food groups are small or close to zero. Only dairy products and meat could be identified as main substitutes. Analyzing the demand for fish on more disaggregated product categories revealed in highly elastic demand. Most fish products are seen as substitutive goods by the consumers. This essay closes a gap in the literature, because fish demand in Germany has not been analyzed in detail since 1985. The results of the own-price elasticity were used in the sixth study to characterize the demand for fish in simulation runs.

The economic benefits from aquaculture R&D conducted in Germany were estimated in the sixth essay **"Simulating the benefits from aquaculture R&D – the impact of elasticities and spillovers"**. IFPRI's DREAM model, which is based on a standard commodity market model with linear supply and demand function, was used for the simulation and estimation of consumer and producer benefits for the EU-15-countries. Limited knowledge on income and demand elasticities as well as supply elasticities and knowledge spillovers was reflected in a sensitivity analysis. More than 100 scenarios were estimated to analyze the effect of modifications of the four parameters on total producer and consumer surplus. The results show that the total benefits are positive when there are no technology spillovers and only German aquaculture producers adopt the new cost-saving technology. When spillovers from Germany to all other EU-15-countries are introduced the total benefits boost. Modifications of the supply elasticity, demand elasticity as well as the income elasticity show that they have an impact on total benefits. In most cases a more elastic supply leads to increasing total surplus, of which consumers also gain a larger share. More elastic demand leads to a decreasing total surplus and also to a decreasing share of consumers in total surplus. Increasing demand elasticity has also positive effects on total surplus. The simulations also show that the effect of modifications of the spillover coefficients have a much larger impact on total benefits than variations in the elasticities. Knowledge transfer and adoption of new technologies plays an important role. The study indicates that research in aquaculture R&D is beneficial for producers and consumers. Moreover, the transfer of knowledge and technologies is very important to boost total surplus, increase the aquaculture production, and to encourage even more new knowledge and inventions.

Stagnating fisheries supply and an increasing demand for fish offer a high market potential for aquaculture fish. Aquaculture production can be increased by means of R&D. The results of the Delphi study show that investments in the following research areas are most promising: These areas are fish nutrition, fish health, water management and quality management. The DREAM-simulation showed that the location of research does not matter as long as there are spillover effects. But, research results will have a greater impact when they are transmitted to more countries and when the spillover effect is strong. New knowledge or inventions will probably spread faster and reach more researchers and aquaculture producers, when the research facility, which creates the new knowledge, is well embedded in a network as when it is only loosely connected in the network of researchers, research facilities, and producers. Identifying and investing in well embedded researchers or research facilities thus leads to higher expected welfare gains. The simulation runs showed that technology and knowledge spillovers boost total benefits. Fostering knowledge transfer is therefore crucially important to increase the aquaculture production. Aquaculture R&D does not make much sense, when the results are not transferred to the aquaculture producers and to other researchers.

Some future research areas arise directly from the essays in this dissertation.

For the simulation of the welfare effects some parameters had to be set, where not enough information is available. This also includes the demand elasticities of fish for every country. But, much less is known on the producer side in the aquaculture sector. Only a few studies exist, which measures short and long term supply elasticities. Not much is known on the adoption and diffusion of new technologies by aquaculture farmers compared to agriculture. More information is also needed on research lags and adoption lags. How long does it take, until a new technology is created and to which costs? And after what time do aquaculture farmers adopt this technology? And how long does the aquaculture sector profit from such a technology and when is the technology replaced by another technology? Not much is also known on the spillovers of knowledge and technologies between countries, between research areas, between aquaculture sectors, and between several species. More information on all these points could lead to a better understanding of R&D in aquaculture and to a refined estimation of their welfare impacts.

The stock of knowledge in a research area is not constant. Ongoing research in the different areas of aquaculture leads to ever more knowledge on aquaculture in total. New research areas may emerge one would have not imagined some years before. To keep the knowledge on aquaculture up to date and does not miss a trend in research, it is useful to repeat a Delphi study at regular intervals. The Delphi method is a well-recognized instrument for long term prognosis of technological evolution and futures research. A web-based Delphi study has the additional advantage of surveying international experts in a relatively short time and relatively low costs. Universities or research facilities, which want to engage in aquaculture R&D, may establish their research programme based on the results of this Delphi study. The results of the Delphi study would also provide funding agencies and decision makers not to miss a promising research area. Investments in aquaculture R&D could thus lead to new knowledge and developments of new technologies.

The aquaculture industry is blamed for its environmental impact, e.g. eutrophication of water or destruction of mangroves. Ongoing R&D, e.g. on filter techniques, combined with increasing control of the production process could also lead to gain more control on the environmental impact. Further, disease outbreaks, like the infectious salmon anemia virus outbreak in Chile since 2007, may be prevented in the future by developing a vaccine. Thus transmission of the virus to other organisms can be avoided and the losses in fish farms can be reduced. R&D on fish feed could also lower the need of fish meal and oil in the aquaculture industry and could thus lower the pressure on wild fish stocks. These effects of aquaculture R&D are not included in the welfare measurement of this dissertation but it would be challenging to include these effects, too.

The essay on the quality of trade data showed that trade data are inaccurate. However, due to limited alternatives to trade data this data have to be used for analyzing trade. Additionally, more specific trade data, which e.g. separate the fish trade data by the origin of the fish from fisheries or aquaculture, would make the trade analysis more specific, too. Maybe the results of the network analysis would change, if this data were available. The network analysis of international fish trade broadens the knowledge on international markets for fish. Tools of the network analysis offer a way to gain information on the structure of the global network as well as on the position and role of each country. This analysis could be extended by identifying cores within the network. Additionally, the analysis could be narrowed to the important trading countries. The network indices could reveal some new insights which remain hidden when the trade of all countries is examined.

The results of the network analysis revealed that trade with fish increased in the last two decades. One reason for that is the growing world demand for fish. Population and per capita income are growing and changing preferences to healthy food might lead to an increasing demand for fish in the future. Analyzing demand is thus a task which should be done regularly to identify changes in consumer behavior. In the case of demand analysis, estimation with more recent data might lead to new insights on the demand for fish in Germany. Additionally, a separation of fish products originating from fisheries and aquaculture could uncover if the consumers react different to price changes for fish from aquaculture or fisheries, respectively. The use of cross sectional data as well as the use of time series data may lead to problems in the estimation of demand systems. Missing price information and zero observations are the main problems in cross-sectional data. Autocorrelation and unit roots are the statistical problems of using time series data. Conducting real field experiments to measure consumers' reactions may avoid the problems one could have with cross-sectional data or time series data. The control for price offerings in such experiments may reflect the consumer behavior much better than one can find it in cross-sectional data or time series data sets. To allow for a more realistic estimation of R&D-induced benefits, country or region specific elasticity estimates would also be necessary. Recent studies applied different methods on different fish species in different periods. International studies on fish demand are thus hardly comparable.

References

Duarte, C.M., Marbá, N. and Holmer, M. (2007): Rapid domestication of marine species. Science 316: 382-383.

Morgenstern, O. (1950): On the accuracy of economic observations. Princeton University Press, Princeton.

9. German Summary

9 German summary

Die Dissertation besteht aus sechs separaten Essays und beginnt mit einer Übersicht der Fischproduktion und des Fischkonsums. Die folgenden Essays handeln von der aktuellen Stärke und der zukünftigen Entwicklung der Aquakultur-F&E, internationalem Fischhandel und der Qualität von Fischhandelsdaten, Fischnachfrageanalyse und der Schätzung der ökonomischen Wohlfahrtsgewinne für Produzenten und Konsumenten, die durch Aquakultur-F&E entstehen.

Das erste Essay „**Captured and cultured fish for food**" dient als kurze Einleitung. Im Vergleich zu F&E in der Landwirtschaft ist die F&E aquatischer Organismen noch eine junge Disziplin. Vor etwa 2000 Jahren hatten die Menschen bereits 90 Prozent der heute gehaltenen bzw. kultivierten landbasierten Arten domestiziert. Im Gegensatz dazu wurden etwa 97 Prozent der aquatischen Organismen erst seit Beginn des 20. Jahrhunderts domestiziert (Duarte et al. 1997). Die Aquakultur ist in den letzten drei Jahrzehnten immer bedeutender geworden und ist verantwortlich für das Wachstum der globalen Fischproduktion, da die Fischereierträge seit Mitte der 1990er Jahre stagnieren. Mengenmäßig dominieren China und andere asiatische Länder die Aquakulturproduktion. Im globalen Kontext betrachtet, ist die Aquakulturproduktion der Europäischen Union (EU) annähern unbedeutend. Der durchschnittliche Fischkonsum pro Person beträgt etwa 16,4 kg/Jahr und liegt damit unterhalb des durchschnittlichen weltweiten Fischkonsums pro Person von 18,4 kg/Jahr. Die EU-27 ist der größte Importeur von Fisch und die Importwerte und –mengen sind zwischen 2004 und 2009 deutlich gestiegen. Der letzte Abschnitt in diesem Essay analysiert die Entwicklung der Anzahl wissenschaftlicher Publikationen. Wenn die Anzahl der Publikationen als Indikator für die F&E-Aktivität in einem Bereich betrachtet wird, kann man auf einen starken Anstieg der Aquakultur-F&E seit Beginn der 1990er Jahre schließen. Ansteigende F&E-Aktivitäten können zu einem Produktivitätswachstum in der Aquakultur führen, so dass innerhalb weniger Jahrzehnte Aquakulturfisch den Fangfisch ersetzen könnte.

Im zweiten Essay „**Testing the quality of international fish trade data**" wird die Qualität von Fischhandelsdaten für den Zeitraum von 1992 bis 2008 untersucht. Die Qualität von Import- und Exportdaten ist wichtig, da verzerrte Daten zu verzerrten Ergebnissen und somit zu verfälschten Schlussfolgerungen führen können. Zunächst wurde untersucht, ob die ersten Ziffern der Import- und Exportwerte Benfords Gesetz folgen. Die Ergebnisse zeigen, dass die Daten einiger Länder deutlich von Benfords Gesetz abweichen. Für diese Länder kann vermutet werden, dass während des Prozesses der Datenerhebung bis zur Veröffentlichung der Daten Unregelmäßigkeiten aufgetreten sind. Dies bedeutet jedoch nicht zwangsläufig, dass die Daten verfälscht wurden. Für Länder, deren Daten Benfords Gesetz folgen, ist dies auch kein Beweis, dass die Daten frei von Fehlern und Manipulationen sind. Für die drei analysierten Fischprodukte folgen die Import- und Exportwerte Spaniens Benfords Gesetz. Die Exportdaten der USA und die Importdaten Deutschlands weichen dagegen jedoch bei den drei Fischprodukten von Benfords Gesetz ab. Insgesamt konnte jedoch kein systematisches Abweichungsmuster gefunden werden.

In einem zweiten Schritt wurden bilaterale Handelsdaten untersucht. Da die gleiche Transaktion sowohl als Import als auch als Export in den Statistiken festgehalten wird, bietet sich dadurch die Möglichkeiten, Handelsdaten zu vergleichen und zu beurteilen. Allerdings ist unbekannt, welcher Wert der wahre Wert ist und ob dieser überhaupt erfasst werden kann. Prozentuale Differenzen der Import- und Exportwerte wurden berechnet und in vier Klassen eingeteilt, die von Morgenstern (1950) vorgeschlagen wurden. Die Ergebnisse zeigen, dass die Länderpaare die Handelswerte zu hoch oder zu niedrig angeben. Einige Länderpaare konnten identifiziert werden, deren Handelsdaten immer in dieselbe Klasse fallen, also z.B. die Importwerte von beiden Ländern immer zu hoch oder die Exportwerte von beiden zu niedrig angegeben werden, oder beides der Fall ist. Für diese Länderpaare kann eine systematische Verzerrung der Handelsdaten vermutet werden. Wenn die prozentualen Differenzen jedoch in mehr als eine Klasse fallen, wird es wahrscheinlicher, dass die Diskrepanzen zufälliger Natur sind. Die Ergebnisse zeigen, dass die Handelsdaten ungenau sind und dass große Differenzen zwischen den Export- und Importwerten existieren. Je disaggregierter die Daten betrachtet werden, desto größer ist der

Anteil der Abweichungen von mehr als ±50 Prozent. Für aggregierte Produkte scheint die Konsistenz der Daten höher zu sein, als es für disaggregierte Produkte der Fall ist. Systematische Abweichungen konnten jedoch nicht entdeckt werden, so dass es jeweils notwendig ist, sich die Daten eines Länderpaares genauer anzusehen. Handelsdaten sollten auf ihre Qualität hin untersucht werden, bevor sie für weitere Analysen verwendet werden. Die Maßnahmen zur Steigerung der Datenqualität sollten von den Statistikämtern fortgeführt werden, um die Qualität und die Konsistenz der Handelsdaten zu erhöhen.

Im dritten Essay mit dem Titel **„Fish in the network – network analysis of international fish trade"** wird der internationale Handel mit Dorsch, Shrimps und Salmoniden analysiert. Methoden der Netzwerkanalyse werden angewendet, um Informationen über die Struktur der globalen Handelsnetzwerke zu erhalten und um die Rolle der einzelnen Länder im Netzwerk zu charakterisieren. Die wichtigsten Länder gemessen am Handelswert und an der Anzahl der Handelspartner werden identifiziert. Die globalen Netzwerkmaße werden für den Zeitraum von 1990 bis 2009 berechnet. Zusätzlich werden die wichtigsten Zwischenhändler der drei Fischarten ermittelt. Die Ergebnisse zeigen, dass der internationale Handel von Dorsch, Shrimps und Salmoniden sowohl vom Handelswert als auch von der Anzahl der teilnehmenden Länder zugenommen hat. Die Anzahl der Handelsverbindungen ist dabei für Shrimps und Salmoniden deutlich höher als für Dorsch, was auf eine stärkere Vernetzung der Länder im Handel mit Shrimps und Salmoniden hinweist. Einige Netzwerkmaße deuten darauf hin, dass der Export sich auf einige wenige Länder konzentriert. Zudem steigt auch die Anzahl der Handelspartner für die Exportländer. Die Bedeutung der Zwischenhändler ist für den Handel mit Dorsch und Lachs rückläufig, steigt jedoch für den Handel mit Shrimps etwas an. Die Ergebnisse zeigen auch, dass sich die Netzwerke in Bezug auf die Herkunft der Fische, also Fischerei oder Aquakultur, unterscheiden.

Das vierte Essay **„The shape of future aquaculture R&D – results of a Delphi study"** fasst die Ergebnisse einer online durchgeführten Delphi-Studie aus dem Jahr 2008 zusammen. Mehr als 270 Aquakultur-Experten haben in der ersten Runde teilgenommen, um den aktuellen Stand der Aquakulturforschung in Hocheinkommensländern zu bewerten. Des Weiteren haben sie auch die zukünftige Entwicklung der Aquakulturforschung bis zum Jahr 2020 eingeschätzt. Insgesamt wurden drei Befragungsrunden durchgeführt. Die Aquakulturexperten sind sich einig darin, dass die Aquakulturforschung bereits viel erreicht hat und dass sich dies auch in Zukunft fortsetzen wird. Sie sind zudem überzeugt, dass die Aquakulturforschung einen starken Einfluss auf die Produktivität der Aquakultur haben wird. Die Forschungsgebiete Fischernährung, Fischzüchtung und Reproduktion sind die Bereiche, in denen bisher die größten Fortschritte erzielt wurden. Die Experten schätzen, dass im Jahr 2020 die Forschungsleistung in den Bereichen Fischgesundheit, Fischernährung sowie Wasser- und Qualitätsmanagement am höchsten ist. Norwegen ist das derzeitig mit Abstand und auch in Zukunft führende Land der Aquakulturforschung. Israel und das Vereinigte Königreich werden auf den Plätzen zwei und drei in Zukunft von Spanien und den USA verdrängt. Die Delphi-Studie zeigt auch die derzeitigen und zukünftigen Forschungsgebiete in den verschiedenen Forschungsbereichen auf. Die Ergebnisse unterstützen Förderorganisationen und Entscheidungsträger die erfolgversprechendsten Forschungsgebiete der Aquakultur zu identifizieren.

Die Nachfrage nach Fisch wurde im fünften Essay „Demand for fish in Germany" untersucht. Auf Basis der Querschnittsdaten der Einkommens- und Verbrauchsstichprobe 2003 in Deutschland wurde ein quadratisches Almost Ideal Demand System (QUAIDS) geschätzt. Zusätzlich wurden Preiselastizitäten von Fisch und Fischprodukten berechnet. Fehlende Preisinformationen und Nullbeobachtungen in den Daten können zu verzerrten Schätzergebnissen führen, so dass zwei Methoden, nämlich ein Ansatz zur Preisbereinigung und ein Verfahren zur Vermeidung des Selektivitätsbias (consistent-two-step estimation), angewendet wurden, um diesen möglichen Problemen zu begegnen. Die Sensitivität der Ergebnisse wurde aber auch im Hinblick auf das gewählte Schätzverfahren ermittelt, in dem das QUAIDS mit oder ohne diese beiden Methoden geschätzt wurde. In der Schätzung mit qualitätsbereinigten Preisen und der consistent-two-step estimation ergab sich eine Eigenpreiselastizität von etwa -0,88 für Fisch. Modifikationen in der Schätzmethode ergaben Eigenpreiselastizitäten im Bereich zwischen -0,25 und -0,96. Das qualitative Ergebnis bleibt jedoch unverändert und die Nachfrage nach Fisch kann in Deutschland als unelastisch angesehen werden. Die Kreuzpreiselastizitäten zwischen Fisch und anderen Nahrungsmittelgruppen sind gering oder annähernd null. Nur Milchprodukte und Fleisch konnten als Substitut zu Fisch identifiziert werden. Die Nachfrageanalyse der disaggregierten Fischprodukte ergab elastische Nachfragen. Des Weiteren werden die meisten Fischprodukte als Substitute von den Konsumenten betrachtet. Dieses Essay schließt zudem eine Lücke in der Literatur, da die Nachfrage nach Fisch in Deutschland seit 1985 nicht detailliert untersucht wurde. Die Ergebnisse der Eigenpreiselastizität wurden in der sechsten Studie verwendet, um die Nachfragekurven in den Simulationen zu charakterisieren.

Der ökonomische Wohlfahrtsgewinn durch in Deutschland durchgeführte Aquakultur-F&E wurde im sechsten Essay „Simulating the benefits from aquaculture R&D – the impact of elasticities and spillovers" geschätzt. Für die Schätzung der Wohlfahrtseffekte für die Konsumenten und Produzenten in den EU-15-Ländern wurde das vom IFPRI bereitgestellte DREAM-Modell verwendet, das auf einem Standard-Gütermarktmodell mit linearen Angebots- und Nachfragekurven basiert. Das begrenzte Wissen über Einkommens- und Nachfrageelastizitäten aber auch über die Angebotselastizität und die Übertragungseffekte von Wissen wurde in einer Sensitivitätsanalyse berücksichtigt. Mehr als 100 Szenarien wurden berechnet, um den Einfluss der vier o.g. Parameter auf die gesamte Produzenten- und Konsumentenrente zu ermitteln. Die Ergebnisse zeigen, dass bereits positive Wohlfahrtseffekte entstehen, wenn es keine Übertragungseffekte gibt und nur die deutschen Aquakulturproduzenten die neue, kostensenkende Technologie einsetzen. Werden Übertragungseffekte von Deutschland zu den restlichen EU-15-Ländern berücksichtigt, steigen die Wohlfahrtseffekte stark an. Änderungen in der Modellierung der Angebots-, Nachfrage- und Einkommenselastizität zeigen, dass sie die Wohlfahrt beeinflussen. In den meisten Fällen führt eine elastischere Angebotselastizität zu einem höheren Wohlfahrtsgewinn und der Konsumenten-Anteil an dem gesamten Gewinn steigt zudem. Eine elastischere Fischnachfrage führt dagegen zu einem Rückgang des gesamten Wohlfahrtsgewinns sowie einem sinkenden Konsumenten-Anteil an dem gesamten Gewinn. Eine höhere Einkommenselastizität hat ebenfalls einen positiven Effekt auf den Wohlfahrtsgewinn. Die Simulationen zeigen aber auch, dass Änderungen in den Übertragungskoeffizienten einen wesentlich stärkeren Einfluss auf die Wohlfahrtseffekte haben, als für die Änderungen der Elastizitätsparameter der Fall ist. Der Transfer von Wissen und neuen Technologien spielt daher eine bedeutende Rolle. Diese Studie zeigt, dass sowohl Produzenten als auch Konsumenten von Aquakultur-F&E profitieren können. Der Transfer von Wissen und Technologien ist entscheidend, um die gesamte Wohlfahrt zu erhöhen, die Aquakulturproduktion zu steigern und um weiteres Wissen und Erfindungen zu fördern.

Literaturverzeichnis

Duarte, C.M., Marbá, N. and Holmer, M. (2007): Rapid domestication of marine species. Science 316: 382-383.

Morgenstern, O. (1950): On the accuracy of economic observations. Princeton University Press, Princeton.